Lifeboat Design and Development No.9

STEAM
LIFEBOATS

The RNLI's steam lifeboats, their design, history and careers

Nicholas Leach
FOXGLOVE PUBLISHING

First published 2023

Published by
Foxglove Publishing Ltd
Foxglove House, Shute Hill,
Lichfield WS13 8DB
United Kingdom
Tel 07940 905046

© Nicholas Leach 2023

The right of the Author to be identified as the Author of this work has been asserted in accordance with the Copyrights, Designs and Patents Act 1988.

All rights reserved. No part of this book may be reprinted or reproduced or utilised in any form or by any electronic, mechanical or other means, now known or hereafter invented, including photocopying and recording, or in any information storage or retrieval system, without the permission in writing from the Publishers. British Library Cataloguing in Publication Data.

ISBN 9781909540286

Typesetting/layout by
Nicholas Leach/
Foxglove Publishing

◀ (page 1) Duke of Northumberland, the RNLI's first steam lifeboat, on the Mersey circa 1893. She was stationed at New Brighton, the station that guards the entrance to the river.

LIFEBOAT BOOKS PUBLISHED BY FOXGLOVE PUBLISHING

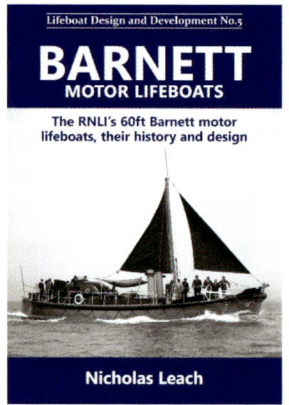

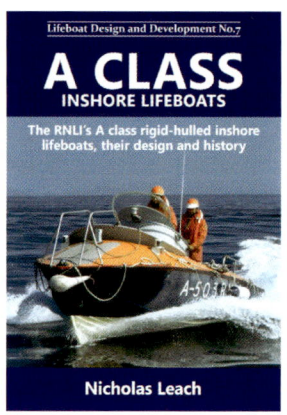
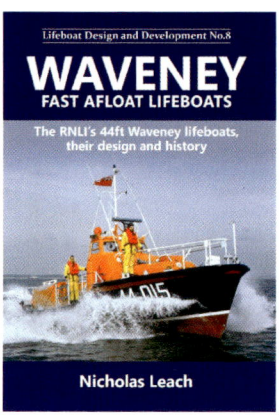

LIFEBOAT DESIGN AND DEVELOPMENT SERIES This is the ninth book in a series of concise illustrated volumes that trace the history of and describe technical aspects of RNLI motor lifeboat types. The other volumes in the series are as follows: No.1 Clyde class rescue cruisers; No.2 Surf motor lifeboats, of which ten were built; No.3 Atlantic rigid-inflatable inshore lifeboats; No.4 47ft Tyne fast slipway lifeboats; No.5 60ft Barnett twin-screw motor lifeboats; No.6 33ft Brede intermediate lifeboats; No.7 A class rigid-hulled inshore lifeboats; and No.8 Waveney fast afloat lifeboats.

THE AUTHOR Nicholas Leach has a long-standing interest in lifeboats and the lifeboat service. He has written many articles, books and papers on the subject, including a history of the origins of the lifeboat service; a comprehensive record of the RNLI's lifeboat stations in 1999, the organisation's 175th anniversary; RNLI Motor Lifeboats, a detailed history of the development of powered lifeboats; and numerous station histories, including ones covering the stations of Cromer, Longhope, Padstow, Sennen Cove, Weymouth and Humber. He has visited all of the lifeboat stations in the UK and Ireland, past and present, and is Editor of Ships Monthly, the international shipping magazine, and Lifeboats Past & Present, the magazine for lifeboat enthusiasts.

Contents

Introduction	5
The development of steam ships and tugs	7
Applying steam power to rescue work	15
The first steam lifeboat	23
Two more steam lifeboats	41
Screw-propelled steam lifeboats	51
Careers of the Steam lifeboats	73
Steam lifeboats round the world	88
Steam lifeboats in the Netherlands	92
Map of stations served by steam lifeboats	96

Acknowledgements

A number of people have assisted with this publication: Hayley Whiting of the RNLI Heritage Trust provided research facilities and much assistance; Iain Booth, Ken Brand and John Harrop supplied illustrations which have helped in recording the history and development of the steam lifeboats. Kees Brinkman at the KNRM in the Netherlands provided photos and information about his country's steam lifeboats, and Richie Leonard supplied some extra information. Ian Moignard thoroughly checked text and contents, and I am extremely grateful to him. Finally, my gratitude extends to pioneering lifeboat historian the late Grahame Farr, of Portishead, whose research into the history of steam lifeboats and his book on the subject, which was published in the early 1980s, were of much use in compiling this volume.

Nicholas Leach
Lichfield, January 2023

Summary of steam-powered lifeboats and tugs

ON	Year (Yd no) Builder	Len Bre	Name Donor	Engines	Stations (launches/lives saved) Disposal
231	1889 (G 227) R. & H. Green Blackwall	50' 14'3"	**Duke of Northumberland** RNLI Funds.	1x170hp water pump, single screw	Harwich 1890-92 (15/33) Holyhead 1892-93 (6/9) New Brighton 1893-97 (29/14) Holyhead 1897-1922 (125/239) Sold 1.1923
362	1894 (G 289) R. & H. Green Blackwall	53' 16'	**City of Glasgow** City of Glasgow Lifeboat Fund.	1x200hp scoop, single screw	Harwich No.2 1894-97 (23/4) Gorleston 1897-98 (1/0) Harwich 1898-1901 (5/28) Sold 10.1901
404	1897 (TH325) J. I. Thornycroft Chiswick	55' 16'6"	**Queen** RNLI Funds.	1x198hp water pump, single screw	New Brighton No.2 1897-1923 (81/196) Sold 4.1924
420	1898 (W 1054) J. S. White, Cowes	56'6" 15'9"	**James Stevens No.3** Legacy of James Stevens, Birmingham.	1x180hp compound, single screw	Grimsby 1898-1903 (6/0) Gorleston 1903-08 (37/30) Angle 1908-15 (12/5) Totland 1915-19 (6/0) Dover 1919-22 (5/1) Holyhead 1922-28 (20/18) Sold 1.1929
421	1899 (W 1055) J. S. White, Cowes	56'6" 15'9"	**James Stevens No.4** Legacy of James Stevens, Birmingham.	1x180hp compound, single screw	Padstow 1899-1900 (4/9) Capsized and wrecked on service, 11.4.1900, eight lost
446	1901 (W 1101) J. S. White, Cowes	56'6" 15'9"	**City of Glasgow** City of Glasgow Lifeboat Fund.	1x180hp compound, single screw	Harwich 1901-1917 (98/87) Sold 12.1900 to Admiralty
478	1901 Ramage & Ferguson	95'6" 19'6"	**Helen Peele** Legacy of Mr C. J. Peele, Chertsey.	2x331hp compound, twin screw	Padstow 1901-29 (working in conjunction with pulling & sailing lifeboat, 19/10 on own account) Sold 1.5.1929

Bibliography

Benham, Hervey (1980): The Salvagers (Essex County Newspapers).

Beukema, Hans (2005): S.O.S. Voor de Hoek (Maritext vof, Delfzijl).

Farr, Grahame (1981): Papers on Life-boat History, No.5: The Steam Life-boats 1889-1928 (Grahame Farr, Bristol).

Green, J.F. (1891): Steam Lifeboats (Paper read before Society of Arts, 14 January 1891).

Harrison, Eric (1970): Cross-Channel Curiosity (Sea Breezes, Vol.44, pp.601-603, Oct 1970).

Leach, Nicholas (2005): RNLI Motor Lifeboats (Landmark Publishing Limited, Ashbourne).

— (2011): Harwich Lifeboats: An Illustrated History (Amberley Publishing, Stroud, Gloucestershire)

— (2014): Holyhead Lifeboats: An Illustrated History(Foxglove Publishing, Lichfield).

— (2015): New Brighton Lifeboats: An Illustrated History of 150 Years of Life-saving on the Mersey (Foxglove Publishing, Lichfield).

Malster, Robert (1974): Saved from the Sea: The story of life-saving services on the East Anglian coast (Terence Dalton, Lavenham, Suffolk).

Introduction

Steam-powered lifeboats were introduced by the Royal National Lifeboat Institution (RNLI) to the lifeboat fleet towards the end of nineteenth century, and were the first powered lifeboats to take up operational duties in the British Isles. Only six steam-powered rescue craft were built, together with one steam tug, but almost all gave fine service at key stations. The main benefits of powered lifeboats over those relying on sails and oars, or a combination of the latter, were: (1) the ability to head into wind and sea; (2) the ability to operate further afield, reaching casualties beyond the relatively limited range of pulling lifeboats; and (3) the ability to reach casualties faster.

The introduction of steam power to the lifeboat fleet came relatively late. Steam had been first used to power ships more than half a century before the RNLI started seriously considering its application to rescue work. This was due to a number of factors, a primary issue being the technology of the era, which was fairly basic. The issues to be overcome included constructing a steel hull strong enough, installing a steam

◄ The first steam lifeboat, Duke of Northumberland (ON.231), at Holyhead. Named after the President of the RNLI, she was powered by an early form of waterjets.

▲ A generic coloured line drawing of a steam lifeboat, similar to James Stevens No.3 and No.4.

engine of sufficient power and size to fit inside a boat and be able to propel it, and the method used to drive the boat through the water in terms of propellers or waterjets, both of which were used in steam lifeboats.

As well as these technical issues, manning the boats was also a major consideration. Lifeboat crews are almost exclusively made up of volunteers, but on the steam lifeboats the RNLI employed several men to not only take charge of the boat but also maintain the engines and boilers, and ensure their smooth running when the boat was at sea. In rough seas it must have been extremely challenging to keep the fires stoked and pressure high down in the cramped confines of the engine room, especially when the boat rolling and pitching, but was clearly essential to ensure the boat maintained speed and power. The crews on deck must also have suffered when the weather was bad, as there was little or no shelter or crew protection, unlike on modern lifeboats.

The end of the steam era came in 1929, and in the September 1929 edition of The Lifeboat an article entitled 'The End of Steam' reported the removal of the steam tug Helen Peele, stating: 'for thirty-eight years there were Steam Life-boats in the Institution's Fleet. They went out on service 468 times. They rescued 673 lives. These figures show the value of their work, restricted though it was, in the years before the coming of the internal-combustion engine.' A fine record indeed and testament to the builders and crews of the RNLI's steam lifeboats.

The development of steam-powered ships and tugs

The Royal National Lifeboat Institution (RNLI) first designed and built steam lifeboats towards the end of the nineteenth century. The six such craft built by the charity were by far the largest and most powerful lifeboats of their era. The first steam lifeboat to enter service, Duke of Northumberland, was sent to Harwich, in Essex, in September 1890, its introduction coming more than half a century after the first proposals for applying steam power to rescue work had been made.

Steam-powered ships for ocean travel came into operation during the early 1800s, and during the century their use became increasingly widespread. Various technological advances meant they were reliable enough to be of practical use. The first commercially successful steamship service in Europe was operated by the paddle steamer Comet, built for Henry Bell, hotel and baths owner in Helensburgh. In August 1812 Comet carried paying passengers between Glasgow and Greenock, enabling the Clyde to lay claim to being the birthplace of European steam navigation and steamship services.

◀ The ground-breaking steamship Great Britain fitting out alongside Gas Works Wharf, in the Bristol Floating Harbour, April 1844. This historic photograph by William Fox Talbot is believed to be the first ever taken of a ship.

The first seagoing steam-powered craft was Richard Wright's steamboat Experiment, an ex-French lugger, which steamed from Leeds to Yarmouth in July 1813. The first iron steamship was the 116-ton Aaron Manby, built in 1821 by Aaron Manby at the Horseley Ironworks. She became the first iron-built vessel to put to sea when she crossed the English Channel in June 1822, transiting as far as Paris.

Paddlewheels were the main form of propulsion used on the early vessels and, while they were effective in benign conditions, when conditions became more challenging they presented serious drawbacks. As the paddle-wheel performed best when it operated at a certain depth, when the draught of the ship changed, for example due to additional weight, a paddle wheel became too submerged which substantially diminished its performance. But as steam technology advanced, steam engines were mounted in ever larger vessels, and screw propellers were fitted in place of paddle wheels, making ships faster and more economical, and giving them a greater range.

The first steamship purpose-built for scheduled transatlantic crossings was the side-wheel paddle steamer Great Western, designed in 1838 by the famous engineer Isambard Kingdom Brunel. She was superseded by Brunel's next great steamer, Great Britain, built for the Great Western Steamship Company's transatlantic service between Bristol and New York City. While other ships had been built of iron or equipped with a screw propeller, Great Britain was the first large ocean-going ship in which these features were combined, and in 1845 she became the first iron steamer to cross the Atlantic Ocean. She had an long and unusual career, and is now restored and on public display in Bristol.

As steam power was being used in an increasingly wide range of vessels, steam tugs were developed and built. These small but powerful craft were specifically designed to undertake towing operations around ports, assisting ships in and out of harbours, rivers and estuaries, and were soon involved in salvage work. The first steam-powered tugs were paddle driven, but designers soon began experimenting with more effective forms of propulsion. Screw propellers were adopted in the 1870s, as they provided much more power than paddlewheels and enabled the tugs to be more manoeuvrable, while iron and steel was adopted for the hulls, enabling ever more powerful tugboats to be

◀ The steam tug Vulcan at Ramsgate. An iron-hulled steam paddle tug of 140 tons, she was built at Blackwall and delivered to Ramsgate in 1858. In January 1881 she and another of the Ramsgate tugs, Aid, worked with the sailing lifeboat Bradford during the famous rescue of the barque Indian Chief. (By courtesy of the RNLI)

constructed. Companies then started building even larger tugs and expanding into ocean salvage work.

The use of steam tugs to tow lifeboats on rescue work became a common occurrence at a number of the major ports around the country, notably Ramsgate, Gorleston and Harwich on the east coast, and Holyhead in Anglesey. An arrangement was reached with tug operators whereby pulling and sailing lifeboats would work in conjunction with the steam tug, which would tow the boat to the scene of a casualty. Unlike the sailing lifeboats, steam tugs could, of course, head into open waters against an unfavourable wind, or operate in dead calm conditions. Many coxswains and crews regarded this as an ideal arrangement, as the close-quarters manoeuvrability of a pulling lifeboat when attending a wreck was better than that of a larger and more cumbersome steam tug, but the tug itself was ideal in getting to a casualty, particularly when greater distances had to be covered.

Some of the more notable steam tugs were those in use at Ramsgate in Kent, where several were operated in and around the harbour during the nineteenth century. The harbour's development and expansion coincided with the growing use of such vessels and Ramsgate proved to be an ideal location for the craft, with the local boatmen undertaking a considerable amount of salvage and towage work in the approaches to the busy Thames Estuary and the Port of London. Many large ships in distress were assisted by the tugs, while the salvage of wrecks became an intense and contested undertaking, offering substantial monetary rewards to the tugmen, who were often poorly paid.

Ramsgate's tugs were a common sight in the harbour and were kept, according to a contemporary account, with 'crew on board and steam up, ready to put to sea at a moment's notice'. The first Ramsgate tug, built of wood and measuring ninety tons, was named Samson. Built at South Shields by Woodhouse in 1843, she was powered by a 50hp engine, remaining in use until the 1850s. Another wooden-hulled paddle tug, Aid, measuring 89ft by 18ft, of 112 gross tons, was built at Blackwall on the Thames, and was operated out of Ramsgate between 1855 and 1890. Vulcan, an iron steam paddle tug of 140 tons and measuring over 140ft in length, also built at Blackwall, was delivered to Ramsgate in 1858, and Fabia, which was still in service as late as the Second World War, participated in many rescues working alongside the local lifeboat crew. Two further tugs named Aid were operated at Ramsgate: the first was a steel-hulled vessel of 126ft in length and was in use from 1890 to 1915; the second saw service from 1914 to 1925.

The famous Indian Chief rescue

Steam tugs were involved in some truly remarkable rescues, none more so than that in January 1881 when Vulcan and Aid worked with the

▲ A painting of the wreck of the barque Indian Chief, with the Ramsgate lifeboat Bradford in attendance, as well as the steam tug Vulcan which towed the lifeboat to and from the scene in heavy broken seas, January 1881. (By courtesy of the RNLI)

Ramsgate lifeboat Bradford, a 44ft twelve-oared self-righting type craft, on service to the barque Indian Chief, on what became one of the most famous rescues in the history of the RNLI. The 1,238-ton barque was only four days into a journey from Middlesbrough to Yokohama when a north-easterly gale drove her onto Long Sand, off Clacton-on-Sea, at the mouth of the Thames. The crew lit beacon fires and rockets to summon help, and then sheltered from the intensifying storm in the cabins.

Indian Chief was mercilessly pounded by the waves, and the crew's attempts to abandon ship in small boats failed when the boats were engulfed by heavy seas as soon as they were launched. By the end of the afternoon the ship was almost completely under water and the remaining crew on board were forced to climb into the rigging and lash themselves to the spars. All night the ship was battered, with a huge wave bringing down the main and mizzen masts and all who clung to them, leaving only the foremast standing, with eleven of the crew clinging to it for dear life.

Meanwhile, the lifeboat Bradford was towed out from Ramsgate towards the wreck by Vulcan. Unable to find the stricken ship in the darkness, the lifeboatmen spent an uncomfortable night huddled together as the two vessels were tossed about in the storm. As day broke, they spotted what was left of Indian Chief and made for it, facing

STEAM LIFEBOATS

tumultuous seas. Making fast to the stricken ship with a hawser, Bradford took on board the eleven survivors from the foremast and one from the mizzen mast (who died before the lifeboat reached home). The rest of the twenty-nine-man crew had drowned. It took several more hours in terrible conditions before the lifeboat arrived back at her station.

Following the truly remarkable rescue, Ramsgate harbour master Captain Braine wrote: 'Of all the meritorious services performed by the Ramsgate Tug and Life-boat, I consider this one of the best. The decision the coxswain and crew arrived at to remain till daylight, which was in effect to continue for fourteen hours cruising [sic] with the sea continually breaking over them in a heavy gale and tremendous sea, proves, I consider, their gallantry and determination to do their duty.' It is unlikely that such an extraordinary rescue and feat of endurance could have been accomplished without the assistance of the steam tug.

Several gallantry awards were made to the Ramsgate lifeboat crew in recognition of their considerable efforts during this rescue, with Coxswain Charles Fish being awarded the RNLI Gold medal. Silver medals were presented not only to the eleven lifeboat crew, but also to the master of the tug Vulcan, Alfred Page, and the tug's six crew. Notably, lifeboats from Aldeburgh and Clacton also proceeded to the scene but, without the aid of steam power, were unable to reach the wreck. The Harwich lifeboat Springwell was towed to the Long Sand by the paddle tug Harwich, but the owner of the tug, who was on board his vessel as they made for the scene, decided that there was too much risk and ordered

▶ A painting, by Frederick Tudgay, showing one of the Ramsgate tugs towing the lifeboat. The use of tugs for lifesaving work was a common occurrence at the major ports round the British Isles in the nineteenth century.

◀ The steam tug Victoria and the Magazineslifeboat working together to rescue passengers from the packet ship St Andrew, of New York, off Liverpool, on 8 January 1839. The hurricane of 6-7 January 1839 crossed Ireland, North Wales, Liverpool and proceeded east towards Yorkshire, with the wind speed at Liverpool exceeding 100mph. The Magazines lifeboat was operated by the Liverpool Dock Trust and was the forerunner of the New Brighton station, covering the entrance to the River Mersey. (By courtesy of the RNLI)

his skipper to turn for home. The coxswain of the lifeboat realised that without the tug's assistance the lifeboat could not effect a rescue.

Although the steam-powered tugs involved in rescue work at Ramsgate were perhaps the best known, probably because of the widespread publicity given to the Indian Chief rescue, tugs often worked in conjunction with pulling and sailing lifeboats at other major ports during the second half of the nineteenth century: not only Harwich, but also Gorleston, Holyhead and Liverpool, with much success.

One of the earliest rescues in which a steam tug was involved took place off the coast of Anglesey in North Wales on 4 January 1852 almost thirty years before the Indian Chief rescue, when a steam tug assisted with the rescue of the steamer Town, of Wexford, which was stranded on Clipera Point. Because of the severity of the gale, the lifeboat was towed to the scene by the Chester & Holyhead Railway Company's steam tug Anglia, and in three trips managed to save forty-three passengers from the wreck; half an hour after the last survivors had been landed, the steamer broke up and sunk. As well as the usual monetary rewards being made for this rescue, which involved the largest number of survivors being saved by this lifeboat in a single service, the Thanks of the RNLI's committee on vellum were sent to Captain Thomas Hirst, Marine Superintendent of the Chester and Holyhead Railway Company, 'for his promptitude in getting the steam up in the Company's steamer, and towing the lifeboat across the Bay to the wreck'.

STEAM LIFEBOATS

◀ The steam tug Merrimac, in the Pound at Harwich, was used in conjunction with the pulling and sailing lifeboats for many rescues. She was owned by Pauls, merchants of Ipswich, and was active in salvage work as well as being used for towage and summer pleasure trips. (By courtesy of Ken Brand)

On the Norfolk coast, the Gorleston lifeboat undertook many missions in conjunction with the local steam tugs, and with more than one company owning tugs at Great Yarmouth there were, in theory, always tugs available to take the lifeboats to sea. On occasion, however, the tug owners refused to risk allowing their vessels to leave harbour in bad weather with the lifeboats in tow, at which point locals questioned why a steam lifeboat, operated independently of private tug owners, was not available. This was undoubtedly a factor in determining the provision of a steam lifeboat for the station in the 1890s.

At Harwich, when the RNLI first established a lifeboat station in 1876, the owner of the steam tug Liverpool, which had already been instrumental in saving many lives of the Essex coast, said that his vessel would be available to tow the lifeboat to incidents free of charge. A number of steam tugs were operated from Harwich, including Merimac and Spray, as well as the 123-ton wooden-hulled Harwich, which had a long career on the east coast mostly undertaking salvage work. The pulling and sailing lifeboats operating in and around the Thames Estuary, such as from Harwich and Clacton, often had to travel considerable distances to reach a wreck, and a steam tug made a significant difference in enabling the lifeboat to operate at a far greater range than it could have on its own. And it was evident that at these stations a steam lifeboat, with its greater range, would also be an ideal option to perform such rescue work.

Applying steam power to rescue work

The development of the steam lifeboat at the end of the nineteenth century was, in the context of the development of steam-powered ships, somewhat tardy. The seeming reluctance of the RNLI to design, build and operate a steam lifeboat is difficult to explain, particularly given the success of the steam tug when working with lifeboats, and the fact that a number of proposals for applying steam power to life-saving had been made earlier in the century, the first from Sir William Hillary, of Douglas in the Isle of Man, better known for hi sinvolvement in the founding of the RNLI in 1824. The year before the Institution came into being, Hillary had written and published 'An Appeal to the British Nation on the Humanity and Policy of forming a National Institution for the Preservation of Lives and Property from Shipwreck'. In this, he set out his ideas for forming a national body whose main responsibility would be the preservation of human life from shipwreck, having been prompted to act after realising the terrible loss of life that was occurring in the seas surrounding the British Isles.

Sir William can also be credited with the first proposal for a steam lifeboat. His booklet, written in the somewhat verbose style of his age, explained that: 'The principle I propose is to combine the safety and the incapability of being submerged, which the lifeboat possesses, with the commanding power of being impelled against both the wind and a heavy sea, which the steam vessel alone can effect to any great extent'. The steam lifeboat was described in three of the pamphlet's eight pages.

Hillary realised the importance of keeping the boat small enough for useful rescue work, while it had to be large enough to take a steam plant with a suitable power to weight ratio. He suggested that 40ft would be a good length for such a craft, but stressed the need for a wide beam to make it stable in heavy seas. The wide beam, shallow draught and 'low construction' should also help the vessel take the ground if necessary.

For buoyancy, he envisaged cork filling every compartment to the waterline, except the engine room, which needed to be set low in the hull. Compartments all round the sides should be filled with cork rather than being airtight, to withstand any violent shocks, but the boiler should be kept above the waterline. He envisaged a boat which was mainly open, with a half-deck forward and a deck over the engine. Any water taken on board would be expelled through valves opening outwards. To cater for all eventualities, he stated that two masts would also be needed and a lug or schooner rig carried, along with a full set of oars.

Steam vessels at the time were propelled by paddle wheels and, to protect the vulnerable wheels, Hillary suggested strong curved timbers running from the vessel's bows to her quarters, supporting the outside of the wheel axles and affording them protection. He explained: 'They [the timbers] would defend the paddles when running alongside a vessel in distress, or near to rocks, to save them from injury at a time when everything depended on them.' Another important feature was the provision of steam-driven pumps for use to extinguish fires, which he rightly considered as a major hazard in the rough seas to which the boat could be subjected. A steam lifeboat could be also used as a pilot boat and that off ports such as Liverpool, for example, such a craft might be of particular benefit. However, nothing came of these proposals, partly because the paddle wheel was too great an encumbrance when such a craft went alongside another vessel in heavy weather, and possibly because of the cost of such a craft.

The next proposal for a steam lifeboat came in 1851 with the Northumberland Prize competition, instigated by the Duke of Northumberland, who invited naval architects and designers to submit plans for a suitable lifeboat at a time when the national lifeboat service was undergoing a period of major reform. Out of the 289 entries received, just one was for a steam lifeboat, and this came from a civil engineer, G. Rimington, of Wandsworth. The model of his craft was exhibited at the Great Exhibition of 1851. The craft, to be powered by a 10hp engine driving a screw propeller, would measure 40ft by 8ft, and it was estimated that construction would cost £500. However, as with Hillary's proposals, nothing came of this, and the Prize was eventually awarded to a design of self-righting pulling lifeboat.

▲ A drawing entitled 'Life-boat in tow' which appeared in the August 1871 edition of The Lifeboat as a generic illustration of the use of steam tugs to tow lifeboats to stranded ships. It appears in the same issue of The Lifeboat in which an article dismissing the use of steam lifeboats was printed.

Steam lifeboats were occasionally proposed between 1860 to 1880, and were briefly noted in The Lifeboat, the journal of the RNLI. One, discussed in committee on 3 August 1865, was from G.W. Watson, of Lower Shadwell, possibly a forebear of George Lennox Watson, who later became famous as the designer of lifeboats and yachts. On 2 May 1867, the committee viewed a model of a steam lifeboat from John Cowell, of Ware, and on 8 January 1870 they discussed a proposition by a Frenchman, M. Rodolphe, sponsored by Peacock Brothers, of Sunderland. However, details of these proposals have not been found, and in any case the RNLI's committee members were reluctant to pursue them, remaining steadfast in their opposition to the use of steam power in lifeboats.

Indeed, this opposition was further explained in the August 1871 edition of The Lifeboat in which an article, entitled 'Steam Life-Boats', assessed the possibilities of applying steam power to rescue work. However, in describing the various difficulties they were then deemed insurmountable, and the use of steam was dismissed. The difficulties detailed in the article, the author of which is not named, included launching a steam boat through heavy surf, when 'extreme difficulty in sufficiently protecting the fires, to prevent their being extinguished' would be experienced, and 'the extremely violent motion to which [life]boats are often subjected' was believed to be too great for a steam lifeboat. The article was largely speculative and no evidence was given

STEAM LIFEBOATS

regarding any trials with steam power, but did acknowledge the benefits of steam tugs, and concluded: 'Impressed with the importance and apparent immobility of the difficulties which we foresee, . . . we do not feel able to expect that steam life-boats will ever come into general use.'

Captain Busk's Steam Life-ship

Although the RNLI showed little interest in steam rescue craft, others did. In 1870 Captain Hans Busk published a booklet entitled 'Steam Life-Ships and how to build them'. Busk was the owner of several yachts, including one which was steam powered, and in 1869 formed a private lifeboat station at Ryde, on the Isle of Wight, where a Lamb and White whaleboat type boat, 28ft 6in by 6ft 3in in size, was stationed, performing useful service for more than thirty years. The Captain's principal claim to fame was founding the Volunteer Army movement in 1858, and the Ryde lifeboat was built from testimonials he received in recognition of this accomplishment. The boat was named Captain Hans Busk after its donor, and lasted until 1905, remaining in service even after the station was taken over by the RNLI in 1894.

Captain Busk's use of the term 'Life-Ship' in his booklet indicated that his plans were for something more than a conventional shore-based lifeboat. His ambitious intention in founding the 'Life-Ship Institute' and advocating life-ship stations involved the provision of, to use his words, 'a number of powerful and buoyant, decked, steam vessels, of 80-100 tons, capable of encountering any weather and commissioned to cruise off those parts of the coast, where wrecks are chronicled with ever-recurring certainty in every gale'.

The suggested boat was to be built on two watertight caissons, each about 120ft long, by 7ft 6in wide and of 5ft internal depth. These were to be strengthened by nine partitions and, in the centre, two or three spaces were to be reserved for bunkers. The caissons were to be held 13ft apart by a strong deck of oak, four inches thick. The resulting craft was about 125ft in length, with a breadth of 30ft, and it would draw 20in of water. A sliding keel, which could be lowered 5ft below the caissons, was held in a central well, and on each side of this keel was a propeller.

On top of this main deck an airtight square tube, with 3ft sides, was to be built all around the vessel to provide additional buoyancy. This tube

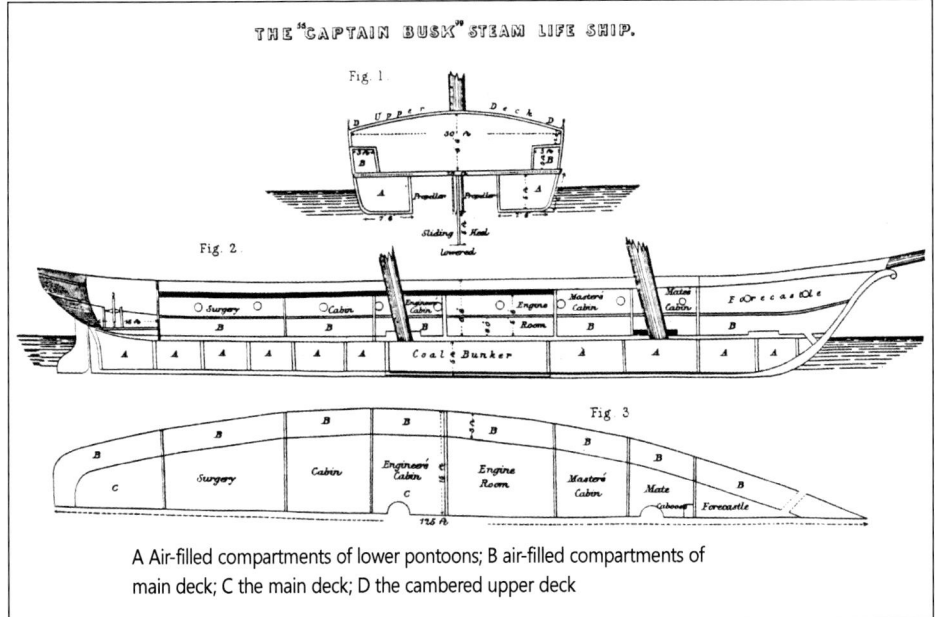

A Air-filled compartments of lower pontoons; B air-filled compartments of main deck; C the main deck; D the cambered upper deck

was to surround an area 22ft wide at deck level and 30ft at upper deck level, where there would be, from forward, a forecastle, mate's cabin, master's cabin, engine room, engineer's cabin, cabin, surgery and lastly, an open cockpit, 15ft in length, containing the wheel. The main deck was to be level, but the upper deck had a foot of camber to facilitate the draining of water. Two hollow masts were to be provided, the after one serving also as a funnel, and the forward one to admit air below decks. Three sails were planned, staysail, foresail and mainsail, with a small jib.

The two propellers, situated almost amidships would be 'working always in smooth water and being all times equally immersed, . . . act uniformly to the greatest possible advantage, and will be quite secure from being fouled by wreckage'. The propellers were to be capable of being raised when not required. They were to be driven by a pair of steam engines, each of 25hp, of an unspecified pattern, which would give, he thought somewhat optimistically, a speed of between twelve and fourteen knots. Later he spoke of using engines of 70-80hp. Captain Busk admitted to being inspired by the yacht Water Spirit, built by William Lawton, of Neston, Cheshire, for Mr Melling, Secretary of the Royal Mersey Yacht Club. She was said to be experimental, of about

▲ Diagram of the Captain Busk Steam Life Ship taken from Busk's book 'Steam Life Ships and How to Build Them', published in 1870 by Sotheran, London.

fifteen to twenty tons, cutter-rigged, and drawing about ten inches of water. One infers she was of a design similar to that of a catamaran.

Successful though the smaller version may have been, it was by no means certain that an enlarged version propelled by steam would perform as well. The Captain thought that each 'life-ship' would cost less than £6,000, but at the time his book was published the Life-Ship Institute had collected just over £700. Faced with major financial challenges, the Life-Ship Institute proposed to build its first craft as a sailing vessel, at a cost of about £1,600, leaving space in the hull for the addition of a steam engine and boiler later.

Somewhat surprisingly, one of the life-ships was actually built. Named Peronelle, she was launched in 1873 at Woolston in Hampshire, and was of forty-three register tons and schooner rigged. Details of the event appeared in the Illustrated London News, which provided information about the financial side of the business but published no technical data. A small engraving, apparently based on one in Busk's book, was used, with the plume of smoke painted out, and it is likely that the vessel had not been fitted out at the time.

On the financial side, the chief supporter was the Greek Consul, Michael Spartali, with the Baroness Burdett-Coutts and the Duke of Wellington among many benefactors donating in cash and in kind. However, Busk died in 1876 and his committee was still unable to finance the installation of an engine in their craft. So they presented the life-ship to the RNLI, which acknowledged the gift at a committee meeting on 2 November 1876, and at once put it up for sale. No records exist to suggest the vessel ever proved herself in a life-saving capacity and clearly she did not fit with any aspect of the RNLI's development programme at the time. The Mercantile Navy List later recorded that she had again been sold, this time to an owner at Port Stanley, in the Falkland Islands, where she was probably employed as a pilot vessel, until she was removed from the registry in about 1895.

Steam power needed

Busk's efforts were undertaken in isolation, although the need for more powerful lifeboats was highlighted by a tragedy in the English Channel off Dungeness in January 1873, when some kind of steam-

▲ An illustration of what the Peronelle Steam Life Ship might have looked like in action. However, she neither received her steam engine nor was she ever used for lifesaving.

powered vessel to supplement the coastal fleet of pulling and sailing lifeboats would have been invaluable. On the night of 22 January 1873 the Spanish steamship Murillo, it was alleged without lights, collided with the Tasmania-bound three-masted ship Northfleet, owned by John Patton of London, which was at anchor two miles offshore among a number of other windbound craft. As well as a cargo of railway iron, the 951 gt Northfleet had 379 persons on board, mostly emigrants heading to Australia, some of whom were rescued by a pilot cutter and a fishing lugger, but 293 were drowned as a result of the collision. The RNLI responded to the disaster by establishing another station at Dungeness in 1874, but as both launching and communications remained a problem at this remote spot, coverage of this part of the coast was only marginally improved. A faster and more capable lifeboat would have helped.

Sir Edward Watkin, Chairman of the South Eastern Railway Company, also responded to the tragedy. He wrote to The Times appealing to the government to establish a life-ship which could cruise off Dungeness, but nothing came of the idea. Just over five years later, in May 1878, a further tragedy involving heavy loss of life occurred near the same place, when the German warships Grosser Kurfürst and Koenig Wilhelm were in collision, and the latter sank with the loss of 287 lives, but still nothing was done to significantly improve safety in the area.

Jubilee Steam Life-Ship No.1

Almost a decade passed before the next advance, which came during 1887 when, thanks to the persistence with which Sir Edward pursued his case, the life-ship proposal took a step forward with the building of Jubilee Steam Life-Ship No.1 for the South Eastern Railway Company. The fact that it was designated 'No.1' speaks boldly for the intentions of her sponsors, but she was destined to remain a one-off. Built by Samuda Brothers, of Poplar, on the Thames, this vessel was 123.5ft long, 20ft in beam, had a 9.1ft depth, and measured 152 gross tons.

She was fitted with twin compound engines of 70hp, built by T.A. Young and Son, of Blackwall, although apparently fitted with a single screw only. On 6 July 1887, while she was on her delivery passage, she achieved a respectable twelve knots. Her lifesaving equipment included lifelines and lifebuoys around her rails, and a 'resuscitation cabin' containing a large bath with hot water laid on. The Director's Report of the Railway Company stated that she was to be stationed at Folkestone and, although specially designed for saving life at sea, would be available for transporting merchandise as well as for towing work.

Jubilee was, in fact, never called upon to fulfil her primary purpose. There were occasions when she might have done so, had she been afloat or at hand and not in Boulogne loading perishable goods. Twelve years later, soon after the formation of the South Eastern and Chatham Railway Co, on 1 January 1899, through the amalgamation of two lines, she was sold to the Shell Transport and Trading Co, who sent her to the East Indies. On 14 September 1901 she was wrecked on a reef in the Celebes Sea.

Although Jubilee never undertook lifesaving work, at the time of her construction the advantages that steam power offered had become clear, and the RNLI went ahead with the construction of a steam lifeboat after a design by the boatbuilder R. & H. Green, of Blackwall, submitted in June 1888, was deemed to be suitable.

▼ The smaller vessel on the left is the Jubilee Steam Life-Ship No.1 (ORN.94294) in this view of the Stade, Folkestone dated about 1890.

The first steam lifeboat

The RNLI's management were opposed to steam power for much of the nineteenth century because no suitable steam plant was available to fit into the 40ft hull they envisaged, despite the fact that tugs had clearly shown its benefits in rescue work. By the time the Institution began seriously considering using steam – almost twenty years after the idea of a steam lifeboat had been completely dismissed, it should be remembered – the technology was a century old. However, matters seem to have taken a new turn during the 1880s and by the end of that decade the first steam lifeboat was undergoing trials. Even with progress in the intervening years, a boat that was ten feet longer – more than 50ft in length – was needed in which to fit the latest steam plant.

Interest in a steam lifeboat was revived in April 1886 when drawings and models of steam lifeboats were shown at the Liverpool International

▼ Duke of Northumberland undergoing trials in the early 1890s shortly after being built. She was constructed on the Thames by Messrs R. & H. Green.

Exhibition, after which the RNLI decided to appoint a special committee to further investigate the subject. By July 1886 the committee had fully investigated a number of designs, both those shown at the Exhibition and from elsewhere. They had also consulted coxswains at those stations where steam tugs were used, and the unanimous opinion of those on the front line had been that, while tugs were a great help in getting a lifeboat to the scene of a casualty, thereafter oars were the best motive power. The result was that the Committee could not recommend the use of steam power in lifeboats themselves.

However, the Committee's views were challenged, with some in the RNLI persisting that steam power could be beneficial. As a result, seven months later it was decided to offer gold and silver medals for drawings or models of a mechanically-propelled lifeboat, and also for methods of applying power to existing designs of lifeboat, including the self-righting type then in widespread use. Ideas were to be submitted by 1 October 1887, and three independent judges were appointed: Sir Frederick Bramwell, FRS, president of the Institution of Civil Engineers; Sir Douglas Murray, of the Board of Trade; and John I. Thornycroft, well-known designer of small craft such as steam launches. However, after

▼ Plans of Duke of Northumberland. As the first steam lifeboat, she was described in some detail in contemporary engineering magazines, in which this drawing was published.

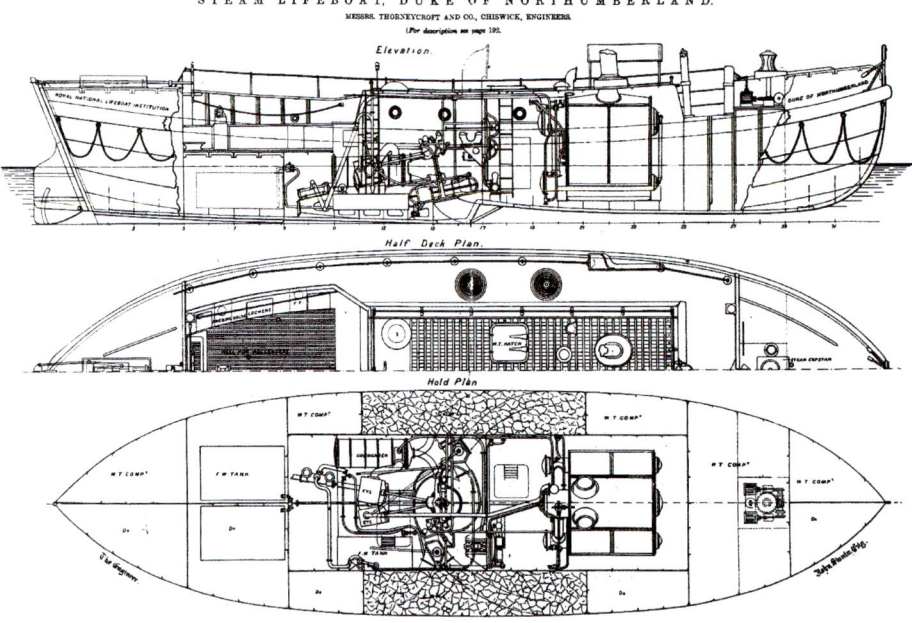

DUKE OF NORTHUMBERLAND'S MACHINERY

The steam engine, built by Thornycroft, of Chiswick on the Thames, was a horizontal direct-acting compound engine with a surface condenser. It developed 170hp and had a stroke of twelve inches, with cylinders of 14½-inch and 18½-inch bore. The boiler was a Thornycroft patent tubular model, with a heating surface of 606 square feet, a grate area of 8½ square feet and a working pressure of 125lbs. Her forced draught fans ran at up to 1,000rpm.

Her water pump was rather a novelty, consisting of an impeller 2ft 6in in diameter on a near-vertical shaft, and delivering water at one ton per minute. The water inlet, in the bottom of the hull, had an area of 1.52 square feet and was fitted with a scoop, and the outlets, placed on each side of the hull, were nine inches in diameter. There were separate forward- and aft-facing outlets, whose water flows were controlled from the cockpit. Bilge keels of elm were fastened on each side of the scoop to afford a measure of protection when going aground.

This propulsion method was revolutionary, being a form of early water jet, which eliminated the need for screw propellers. It was argued that a screw could have been driven by a lighter engine and boiler, because less power would be needed to obtain the same speed and therefore less fuel would be consumed, but the danger of the screw being out of the water while the lifeboat was on service led to the adoption of water jet propulsion. It was also believed that a propeller would be susceptible to being damaged by floating debris or fouled by ropes.

▲ Duke of Northumberland's steam engine and water pump (left), and the boiler (right), which was of the tubular pattern, patented by the Chiswick-based boatbuilder John I. Thornycroft for use in torpedo boats.

considering the proposals, the judges decided that no entry merited an award. Designing and building a steam-powered lifeboat clearly presented lifeboat naval architects with a different set of challenges compared to those of designing pulling and sailing lifeboats.

The special sub-committee remained in existence, however, and, in June 1888, reported on plans for a steam lifeboat submitted by the London-based shipyard R. & H. Green, of Blackwall, long established shipbuilders on the Thames. Green's plans were subsequently modified by joint consultation, found to be suitable, and Green was ordered to build a lifeboat based on them. She was to be named Duke of Northumberland, simultaneously honouring the sixth Duke, then President of the Institution, and the fourth Duke, who in 1851 had effectively rejuvenated the RNLI. Having rejected all proposals for building a steam-powered craft hitherto, the RNLI had, within just a few years, made an about-turn and the use of steam power for rescue work was then pursued over the next decade and more.

A steam lifeboat, operated by the RNLI and thus ready specifically for life-saving duties rather than the towage work primarily undertaken by most steam tugs, offered many advantages over a lifeboat relying on sails, oars or a combination of the two. While doubts remained, even with a prototype under construction, about the efficiency of steam-powered craft for life-saving work, when the new craft was introduced into service any doubts about the efficiency of steam power were soon dispelled.

Duke of Northumberland was launched on 31 May 1889, and measured 50ft in length. She had a moulded breadth of 12ft and an extreme breadth of 14ft 3¾in. Her extreme draught, when she was carrying thirty passengers, all of her crew as well as full bunkers and equipment, was 3ft 6in, and in this condition her displacement was twenty-three tons. She was built of mild steel and strongly riveted, with a third more rivets than were used on a torpedo boat of comparable size. In fact, a total of 72,000 rivets were used in all. Her hull was divided into fifteen watertight compartments.

The sides and deck of the cockpit were fitted with ten large water-relieving valves, and it was surrounded by lockers which formed seats.

▶ Testing the stability of Duke of Northumberland on 26 July 1889 at R. & H. Green's yard at Blackwall on the Thames. During this test, J.F. Green and H.T. Clarke were on board. (Supplied by Leslie Jones)

▲ An etching of Duke of Northumberland which was used in contemporary publications.

Below deck were ballast water tanks with a combined capacity of two tons, which could be emptied by a steam-driven pump according to the number of survivors taken aboard. The decks were covered with corticine (a material made of ground cork and India rubber) to prevent the crew from slipping when they were moving about. A small but powerful capstan was fitted to the forward end box, with its engine in a watertight compartment below, and 120 fathoms of galvanised wire hawser were carried on a reel in the cockpit, adjacent to a strong wire cutter, carried in case the hawser had to be cut in an emergency. The rudder, attached by Hickman's patent fittings, was in two parts. The lower part could be raised and dropped remotely, being controlled by a wheel with self-holding screw gear.

Duke of Northumberland initially spent some months on trials in and around the Thames Estuary, operating out of Sheerness, Southend and Harwich. Her speed trials were undertaken at the Maplin measured mile on 4 August 1890, and on several runs at different trims she averaged a mean speed of 9.175 knots, with her engine achieving a maximum of 450 revolutions per minute (rpm), at which rate coal consumption was only two hundredweight per hour. This became less when she was cruising at about five knots, with her engine achieving 230rpm. These results were deemed excellent and showed that the boat's bunker of three tons of coal gave her a radius of action of 254 miles at just over eight knots.

One of Duke of Northumberland's visits to Harwich was described in the East Anglian Daily Times for 26 July 1890: 'The Royal National Lifeboat Institution has produced the latest novelty in shipbuilding in the shape of a steam lifeboat, and on Thursday afternoon (24 July) this wonderful little craft was on view. She lay in the Thames at Blackwall Pier, and attracted a considerable amount of attention. Visitors were received on board by Charles Dibdin, RNLI Secretary, and Captain The Right Hon H.W. Chetwynd, RN, Chief Inspector of Lifeboats. At three o'clock precisely the order was given to get under way, and the lifeboat at once steamed easily down the river, returning to Blackwall after a very successful run of rather more than a couple of hours. Those who had the good fortune to be on board were charmed with the handiness of the boat and the ingenuity of its construction and machinery.'

During manoeuvring trials, she was found to turn in fifty seconds using the water jets alone, and in forty seconds when both the rudder and water jets were used together. The water jets were found to be more than satisfactory for steering, and this was a distinct benefit should the rudder become damaged. On 12 August 1890 trials were undertaken to assess whether the water intake could be choked by wreckage or weeds and, it was found, in most cases the boat simply passed over the obstacle

▶ Duke of Northumberland at moorings in the Pound while she was on duty at Harwich, the first station to be served by a steam lifeboat. She was at Harwich for less than two years, but saved thirty-three lives in that time. (By courtesy of Ken Brand)

◀ A print showing Duke of Northumberland on service in rough seas by the Cork Lightvessel (off Felixstowe). (By courtesy of Ken Brand)

so the likelihood of the intake becoming clogged was negligible.

Getting the boat ready for a service was initially a concern, but it was found from experience that, starting from cold, it took about twenty minutes to achieve a steam pressure of 100 pounds per square inch, the pressure necessary to get the boat under way. Later, as techniques improved, the boiler was kept warm in bad weather by means of a gas jet from a shore supply and a quicker start was therefore possible.

Operational service

After undertaking further trials during the summer of 1890, Duke of Northumberland was sent for station duties at Harwich in the autumn. She left Blackwall Pier for her designated station on 19 September 1890, stayed overnight at Sheerness and at 12.45pm on 20 September reached Harwich. As steam lifeboats had to be kept afloat, Harwich was an ideal location as its lifeboat was kept at moorings. In addition, most services were to wrecks which occurred on the outlying sandbanks, far enough offshore to be challenging to reach in a pulling and sailing lifeboat, but far easier when using the steam lifeboat.

When Duke of Northumberland was first placed at Harwich, the RNLI stipulated that, whenever she went out on service, she took the sailing lifeboat with her. As a result, her first services involved the then new 38ft self-righting lifeboat Reserve No.3 (ON.206), which was on station temporarily while Harwich's own boat, the 45ft 2in self-righter Springwell (ON.317),

STEAM LIFEBOATS

▲ Duke of Northumberland at Harwich in the early 1890s. She was at the Essex station for almost two years, saving more than thirty lives during that time. (By courtesy of the RNLI)

was undergoing alterations, being fitted with the latest improvements. The practice was not repeated at other stations subsequently served by steam lifeboats. Springwell returned to Harwich in July 1891 and Reserve No.3 was renamed Manchester Unity, being sent to Grimsby where she served from 1893 to 1904, having a period on loan to the Spurn Point station, then operated by Hull Trinity House, from 1901 to 1903.

The Harwich lifeboatmen, under Coxswain Benjamin Dale, were well used to being towed into action, due to an agreement with a local tug company since the RNLI station had been established in 1876. Incidentally, it was the rescue of 173 people from the German emigrant ship Deutschland on 6 December 1875 by the steam tug Liverpool which led to the RNLI placing a lifeboat at Harwich.

When on service, the steam lifeboat's crew consisted of a Coxswain-Superintendent, a Second Coxswain and five crew, while below were the Chief Engineer, Second Engineer and two firemen. William Tyrell was appointed as the steam lifeboat's first Coxswain-Superintendent in 1890, with Arthur H. Simmons the Chief Engineer. Simmons ended up staying with the boat for many years as she moved around the coast.

Duke of Northumberland undertook her first service at Harwich within a month of arriving, on 8 October 1890, when she went to the Cork Sand and stood by the brigantine Ada, of Faversham, which had gone aground. The lifeboats went to the Cork Lightship, and from there found the brigantine, which was bound from Hartlepool for London, with a crew of six. At Ada's master's request, the lifeboats stood by the ship until high water, when the brigantine refloated on the rising tide and was towed off by a commercial steam tug and taken to Harwich.

Less than two weeks later, Duke of Northumberland was in action again, towing Reserve No.3 out at 4am on 20 October 1890 to the 913-ton steamship Achilles, of Sunderland, which was aground on the middle of the Shipwash Sand in a moderate north-westerly gale and heavy seas. As the lifeboat reached the Sands, the lifeboat crew saw a tar-barrel burning on the vessel, and Reserve No.3 was towed alongside. The master, realising his vessel had sustained no damage, asked the lifeboat crews to lighten the ship, so some of the cargo of railway sleepers was thrown overboard, and at high tide the vessel floated off. The steam lifeboat, working with the steam tug Harwich, then towed the steamer clear of the sands, and the vessel resumed her voyage to London. The weekly publication Vanity Fair summed up the benefits of the steam lifeboat after this rescue, saying: 'Apart from doing her own work, the Duke towed one of the Institution's lifeboats to the wreck, arriving there three hours sooner than a boat rowing and sailing could possibly have done'.

In early 1891 Duke of Northumberland completed two effective services, both with the Reserve No.3 lifeboat. The first was on the morning of 6 January, in a north-easterly gale and a very heavy sea. The ketch Day's, of Barrow, carrying scrap iron from London, was wrecked on the Cork Sand. As the lifeboats neared the casualty, the lifeboat crew saw three men in the rigging, two of whom waved for help. Reserve No.3 was towed close to the wreck, and her crew threw grappling irons and lifelines to the two men, who jumped into the water and were pulled into the lifeboat, exhausted and numb. The third man in the rigging was dead. The boats returned to Harwich, where the men were landed and the coxswain of the steam lifeboat decided to return to the wreck to bring ashore the dead body, which had been left in the rigging.

What proved to be the final service by Duke of Northumberland while she was at Harwich came on 3 March 1891, when she went to the three-masted schooner Mercury, of Aberdeen, taking Reserve No.3 with her. The schooner had stranded on the Long Sand in a north-westerly gale, heavy seas and steady rain. The lifeboats reached the Sunk Lightvessel, whose crew confirmed the casualty's position. The lifeboats found the casualty and the steam lifeboat took off the schooner's twelve crew.

On 17 April 1891 the RNLI's Chief Inspector came to Harwich with representatives of the KZHMRS, the South Holland Lifeboat Society, and were taken for a trip in the steam lifeboat. The Dutch visitors requested not to go out to sea, but instead to go upriver to Parkeston as they were later going to be travelling home and wanted to stay dry. During their short trip in calm waters they were reported to be 'much pleased with the boat', while the station's Chief Engineer and Coxswain 'spoke in very high terms of her'. Subsequently the South Holland Society had two steam lifeboats built, the first by Thornycroft at Chiswick.

Although Duke of Northumberland remained on service at Harwich for a further sixteen months after this service, she was undergoing repairs for much of that time. When the Dutch visitors came to Harwich, it was found that the dial plates, used for directing the discharge of water, were unsatisfactory. Further problems were discovered in August 1891 when Duke of Northumberland was hauled up for painting. The heads of bolts securing the intake near the pump had corroded away, probably by galvanic action caused by the copper grating at the intake's mouth. The intake was replaced by a cast steel one of a different pattern, made by Thornycroft, who undertook the necessary repairs at their Chiswick yard on the Thames in September 1891 at a cost of £199 19s 0d.

Soon after these repairs had been completed, the District Inspector discussed with the RNLI's Committee of Management a request from the local committee and Coxswain that the steam lifeboat be allowed to go out on service independently of the sailing lifeboat. The Committee consented to this request in February 1892 with the condition that, when the boat was out on service alone, an additional crew member was taken.

Lifeboat trials at Lowestoft

Before Duke of Northumberland undertook any services alone, however, she was sent to Lowestoft for a series of competitive trials of lifeboats, which took place between 12 February and 19 April. She arrived at Lowestoft on 19 February 1892 after a passage from Harwich during which her boiler's feed water pump broke down, and in the

▲ Duke of Northumberland at Lowestoft during the lifeboat trials in 1892. She did not participate in the trials, so could not be compared with the other lifeboat types, but was instead mainly used for towing the heavy tubular lifeboat. (By courtesy of the RNLI)

trials was mainly used for towing the heavy tubular lifeboat rather than competing. The trials involved assessing four different types of pulling and sailing lifeboat – a Norfolk & Suffolk sailing lifeboat, a Watson sailing lifeboat, a self-righter and the Tubular – to ascertain which best coped with a variety of sea conditions. The tests did not involve assessing the steam lifeboat, other than in her capacity to tow the lifeboats, which she pulled through heavy breaking seas as well as steaming against a tug towing a lifeboat in bad weather, during which the steam lifeboat's performance was considered 'most satisfactory', according to the report in The Lifeboat. The tug achieved a speed of seven and a half knots with the steam lifeboat able to keep up, even though 'the tug [was] being pressed as hard against the sea as prudent'.

After the trials, Duke of Northumberland returned to Harwich. On 25 May 1892 she was taken out of the water to be examined by a Lloyd's surveyor, who recommended various repairs be undertaken. After a further month at Harwich, the RNLI decided that the boat should be assessed elsewhere, so she left Harwich on 22 July 1892 with a record of thirty-three lives saved in less than two years. She was initially sent to Cowes Regatta, where she was opened to visitors, being inspected on 4 August 1892 by HRH The Prince of Wales, with Sir Edward Birkbeck, RNLI Chairman, and Colonel Fitzroy Clayton, Deputy Chairman, in attendance. The Prince was taken for a trip to Osborne Bay, where the boat 'was put through various manoeuvres to show her handiness'.

Anglesey and the Mersey

Following the completion of her royal duties on the south coast, Duke of Northumberland headed to Southampton on 5 August 1892 for some repairs to be undertaken, after which she was transferred to Holyhead, taking up station on Anglesey on 9 October 1892, and serving there until May 1893. As steam lifeboats had to be kept afloat at a sheltered mooring, Holyhead, with its large harbour, was an ideal location. With the steam boat's arrival, Holyhead had three lifeboats, in addition to that at the nearby Porth Ruffydd station, which was operated under the auspices of Holyhead RNLI station, which provided the crew.

Duke of Northumberland undertook six services during her short stay at Holyhead, and in May 1893 was transferred to New Brighton, a busy station at the entrance to the river Mersey. She was sent there on a largely experimental basis and so that 'she may be fully tried under all the varying conditions with which this class of Life-boat has to contend on different parts of the coast'. Although the original intention was for her to be on

▶ Duke of Northumberland at Lowestoft for the lifeboat trials of 1892. She was something of a bystander as the trials focussed on four pulling and sailing lifeboat types: a Watson, a self-righter, a Tubular and a Norfolk & Suffolk. (By courtesy of Ken Brand)

◀ Duke of Northumberland alongside the Mackenzie Pier at Holyhead. The occasion is not known, but it could have been a lifeboat day or publicity event. She spent most of her career at the Anglesey station. (Supplied by Leslie Jones)

the Mersey for a year, she ended up staying for more than four years, launching on service twenty-nine times and saving fourteen lives.

On 22 December 1894, when violent gales swept the country, many lifeboats were called out. In fact, almost every lifeboat from Holyhead to Fleetwood was launched on service, including Duke of Northumberland from New Brighton. She launched at 3.20am and went to the aid of the schooner Faith, of Beaumaris, carrying coal, which was ashore on Crosby Beach in tremendous seas and a severe north-westerly gale. With great skill, Coxswain Martin took the steam lifeboat towards the stranded vessel and, although the lifeboat struck the bottom several times during her approach, she eventually got close enough for the lifeboat crew to be able to rescue the crew of three from the schooner.

The dangers and difficulties of operating a steam lifeboat, with her machinery confined to a relatively small space, were evident during the summer of 1901 when Duke of Northumberland was taken to a boatyard at Birkenhead for an overhaul. After the work had been completed, she was taken out onto the river for trials on 26 June, but during these there was a violent explosion in her boiler room and the two firemen, John Owen and Thomas Owen, were killed. The RNLI's committee was told of the incident: 'while the compass was being adjusted, the engines going dead slow under a pressure of only 56lbs of steam, the steam dome stay broke and was blown clean out, which resulted in the death of the two

STEAM LIFEBOATS

Firemen'. The second engineer, John Hall, who was also in the boiler room, suffered severe scalding, but fortunately escaped alive.

The RNLI's Committee of Management voted £500 to each of the families of the deceased, who left behind widows and a total of twelve children. John Hall, the Second Engineer who had been injured, was awarded £25 compensation by the local committee. Blame for the explosion was attributed to Messrs Thornycroft & Co for a defective piece of equipment in the boiler, with a bad join or weld which, it was determined, was hidden. Although maintaining the steam lifeboats was a costly business, it was clearly essential for safe operations.

Gold medal rescue at Holyhead

After Duke of Northumberland had spent four years on the Mersey, in 1897 the RNLI decided to return her to Holyhead as a new steam lifeboat, Queen (ON.404), had been built for New Brighton. So the original steam boat left New Brighton on 18 December 1897 and the following day arrived at Holyhead, where she stayed until November 1922. Her finest rescue during her time at Holyhead, and indeed one of the most outstanding rescues in the history of the station, took place on 22 February 1908. She put to sea at 12.40pm to the aid of the steamship Bencroy, of Liverpool, which was dragging her anchors in a severe south-westerly gale and

▼ A painting of Duke of Northumberland undertaking the steamship Harold rescue in February 1908. (Painting by Tim Thompson, copy supplied Holyhead RNLI)

◀ Duke of Northumberland berthed in the harbour at Holyhead, with an unidentified steamer berthed close by. (Supplied by Leslie Jones)

very rough seas. When the lifeboat arrived, the steamer was ashore on Holyhead breakwater. Fortunately the tide was rising and, after two of the lifeboat's crew had been put on board to help connect a line from another steamer, Bencroy was towed off to a safe anchorage.

Duke of Northumberland returned to her moorings at about 2pm but, a few minutes later, news was received that the steamship Harold, a small seventy-five-ton vessel, was being driven towards the shore between the North and South Stacks. She was attempting to reach Holyhead while on passage from Teignmouth to Runcorn, when her engine failed five miles from the coast. So the steam lifeboat immediately put out, facing worsening weather with the wind at over eighty miles per hour, conditions in which a pulling and sailing lifeboat would have struggled. But in the mountainous seas, Duke of Northumberland forged ahead with Coxswain William Owen using all his skill, courage and experience to maintain a course as the twenty-eight-ton lifeboat, being thrown about like a cork, fought through the worst seas he had ever encountered.

The steamer's anchor had held when the vessel was close to the rocks, but she was being pounded by enormous breaking seas and the mass of white, foaming water. The crew who worked the boilers below battened down in the stokehold and went through a terrible ordeal during the rescue as, for over two hours, Coxswain Owen fought to get the lifeboat

▶ The Holyhead crew of Duke of Northumberland involved in the rescue of the steamship Harold in February 1908, together with various station officials: back row, left to right, Captain MacKenzie (Chairman), Charles H. Marshall, T. G. Hughes (Treasurer), Samuel Jones, John Lewis, William Owen jnr, Lewis Roberts, McIlgorm (Head Customs Officer), Lewis R. Jones, Captain R.D. Roberts (Lloyd's Agent) and William McLaughlin; front row, right to left (seated): James Lee (Chief Engineer), George Jones, William Owen (Coxswain), Lewis Jones and Hall (Engineer). (By courtesy of Holyhead RNLI)

close enough to save the crew of Harold. Eventually, with outstanding skill and tremendous courage, he manoeuvred the lifeboat close enough for a line to be thrown across to the steamer and seven men were hauled through the churning water, one at a time, to the lifeboat. Suddenly, a huge wave caught the lifeboat and swept her towards the steamer, threatening to smash the two boats together. Desperately, Coxswain Owen fought to avert disaster, but in the moment when the two boats were touching the last two men were snatched from Harold. Owen then called for full speed on the speaking tube to the engine room, and the lifeboat pulled away. The lifeboat, her crew and the survivors faced more heavy seas and had to negotiate the dark coast before the safety of Holyhead was reached. The nine rescued men were taken to the Sailors' Home at 7pm, but their vessel became a total wreck the following day.

Throughout the operation the seas were described as 'mountainous'. The coolness and daring shown by Coxswain Owen and his crew were the subject of universal admiration, and The Lifeboat's account concluded that the 'service was attended by the greatest danger, as the lifeboat was at times in imminent peril of being driven against the disabled ship'. The crew's considerable efforts, together with the Coxswain's gallantry and skilful management, were recognised by the RNLI's Committee of Management and, for his truly outstanding courage, skill and seamanship, Coxswain Owen was awarded the Gold

Medal, with each of the engine-room staff and crew members receiving the Silver medal. They were Thomas Brooke, George Jones, Lewis Jones, Richard Jones, Samuel Jones, James Lee, William McLaughlin, Charles Marshall, William Owen jnr and Lewis Roberts.

Following the Gold Medal service, Duke of Northumberland continued to be much in demand, and she answered several calls every year during the remaining fourteen years of her time at Holyhead. She served throughout the First World War, albeit launching mostly to schooners rather than war casualties. She did undertake some services to war casualties, such as on 23 February 1918 after the Royal Fleet Auxiliary tanker Birchleaf, of London, had been torpedoed and shelled by an enemy submarine and was ablaze. The ship was brought into

◀ A fine model of Duke of Northumberland on display at the Holyhead Maritime Museum on Newry Beach. This gives a good impression of the hull form and deck layout, with the cockpit aft and the twin funnels. In bad weather, the crews on board the steam lifeboats, both full-time and voluntary, must have had significant stamina and faced considerable challenges to maintain steam and deal with the heavy seas. (Nicholas Leach)

STEAM LIFEBOATS

harbour by patrol boats, with the lifeboat crew assisting by running a wire hawser from the casualty to a trawler.

The first steam lifeboat ended her career at Holyhead, and indeed her RNLI career, in 1922, when she was replaced by another steam lifeboat. The last services she performed were fairly routine affairs: on 15 January 1922 she assisted the London & North Western Railway ferry Hibernia, which was in difficulty in a strong gale 200 yards from the breakwater, and on 21 October 1922 she stood by the London steamship Jolly Helen in rough seas half a mile north-east of the Skerries.

Duke of Northumberland left Holyhead in November 1922, having answered 131 calls and saved 248 lives during her time. Added to the six launches and nine lives saved during her few months at the station in the early 1890s, and her other life-saving exploits during that decade at Harwich and New Brighton, when she was sold out of service in 1923 her record of service was outstanding: 181 service launches and 304 lives saved in more than thirty-two years of operational service.

She left Holyhead in November 1922 and was taken to the RNLI's storeyard at Limehouse in London, where she was placed on the sale list. She was sold at auction at the end of January 1923 for £70, but little is known of her career form then until the 1990s, apart from the fact that she was used as a workboat on the Mersey. In 1994 the boat's derelict remains were found at Ready Mix Concrete Widnes works on the river bank, where she was slowly being covered in waste concrete. Although the land was subsequently redeveloped, some remains can still be seen on the river bank. A sad end to a revolutionary lifeboat.

▼ Duke of Northumberland in the large harbour at Holyhead, at 'Porth Sach', her permanent mooring at this time, with the breakwater in the background. (From an old postcard supplied by John Harrop)

THE STEAM LIFEBOAT AND BREAKWATER, HOLYHEAD.

Two more steam lifeboats

Following the success of Duke of Northumberland, both on trials and in operational service, the RNLI ordered a second steam lifeboat in 1893. She was commissioned, like the prototype, from R. & H. Green at Blackwall and details of the new lifeboat were published in The Lifeboat of 1 February 1894. She was designed by G.L. Watson, the RNLI's Consulting Naval Architect who went on to design all the subsequent steam lifeboats, having already designed large non-self-righting sailing lifeboats which went into widespread use.

Although similar to Duke of Northumberland in many respects, the second steam boat was slightly larger, at 53ft overall, with a beam of 16ft and a depth of 5ft 6in. The loaded displacement was thirty tons, with a draught of 3ft 3in. She could carry up to forty people, as well as four tons of coal in the bunkers, and half a ton of feed water in the reserve tank. She had Watson's well-known 'Dora' bow, hitherto built only on sailing yachts; one funnel, rather than two; and the steam capstan was moved from the forward end-box to the engine room casing.

The machinery was, according to The Lifeboat, 'totally different [from that in the first steam lifeboat], and the outcome of many and serious consultations between the Committee of the Life-boat Institution, the builders, and Messrs. Penn, the constructors of the machinery'. A Penn patent watertube boiler was installed to drive two centrifugal pumps with horizontal shafts through a compound engine of 200hp. The pumps were driven by a shaft similar to that of a paddle-wheel engine. The danger of having the intakes fouled by debris was obviated by making them flush with the hull plating, in place of the Thornycroft patent scoop used on Duke of Northumberland. The engines operated at between 358rpm and 420rpm, and produced 220-230 indicated horse power, giving a maximum speed of 9.3 knots and an economical (cruising) speed of 7.34 knots.

STEAM LIFEBOATS

This second steam lifeboat was named City of Glasgow, with funds raised by the people of Glasgow being appropriated to the boat. By the time the boat was ordered, most of the cost had already been raised in the city thanks to the fundraising efforts of the Lifeboat Saturday movement. Once completed, the boat visited her namesake city on 16 June 1894 for a Lifeboat Saturday demonstration at which she was officially named by Mrs Bell, wife of the city's Lord Provost. Also in attendance at the ceremony were Colonel Fitzroy Clayton, the RNLI's Deputy Chairman, and Keppel H. Foote, Inspector of the Northern District.

After her naming, City of Glasgow was returned to the builders at Blackwall for completion and 'bringing up in all points to the specification'. With this work complete, she was sent to Harwich, arriving on 7 November 1894. Her construction was reported by the East Anglian Daily Times in April 1894 as part of an overview of general developments within the RNLI at the time: 'The [RNLI's] Committee are sanguine that the new steam lifeboat will quite fulfil their favourable anticipations, as they believe she will possess many material improvements on the old boat, suggested by experience.'

However, their confidence in the new boat was somewhat misplaced, and throughout her time at Harwich City of Glasgow was plagued with problems. In March 1895, while she was returning to station from a service, the tubes of the condensers started leaking and had to be repaired by the Chief Engineer. In June 1895 the crew expressed their dissatisfaction with the boat's trim, which was down by the stern, while the Chief Engineer stated that the after tank could not be used as a reserve for feed water, even though water was needed because of the amount wasted by the inefficiency of the boiler. Alterations were made in light of these comments, and the modifications 'greatly improved the boat', according to the deckhands and engine room staff. However, further problems were found and, on 31 October 1895, the lifeboat's exhaust pipe cracked at a flange as she returned from service necessitating a series of further repairs.

Perhaps the most serious incident occurred in January 1897 after the port pump's sea water suction pipe joint started leaking two hours after City of Glasgow had returned from exercise. This was so serious that the lifeboat had to be taken to London for repairs to be effected.

▲ The Pound at Harwich with the first City of Glasgow at moorings. The single-funnelled craft was designated the No.2 lifeboat from 1894 to 1897 before going to Gorleston. She returned to Harwich in 1898 but was sold in October 1901 after an RNLI career of less than a decade. (By courtesy of Ken Brand)

Although the Chief Engineer made temporary repairs for the passage, when the lifeboat reached Sheerness during the passage, so much water was coming into the engine room that the Deputy Chief Inspector, who was in charge, had to ground her on a mudbank off Sheerness. The Chief Engineer improvised a further repair, with help from dockyard storekeepers at Sheerness, but as mud had leaked in as well as water, it was impossible to use the engines. So the following day, 6 January, she had to be towed to R. & H. Green's yard at Blackwall. The repairs took three months and City of Glasgow did not return to station until 14 April 1897, the work having cost £235.

Less than three months later, on 10 July 1897, there were further problems. A severe leak developed in one of City of Glasgow's steel pipes, flooding the engine room when the lifeboat was undertaking a Lifeboat Saturday demonstration at Ipswich. Although the leak was plugged, the boat had to again go to London for repairs. However, before arriving at Blackwall another pipe developed a serious leak. Repairs by Green's took two months, with City of Glasgow returning to station on 8 September 1897 under the command of the District Inspector. During the thirteen-hour journey the engines actually worked satisfactorily.

From 24 November 1897 to 1 February 1898 City of Glasgow went to Gorleston for trials to ascertain whether that station could operate a

STEAM LIFEBOATS

steam lifeboat. However, on one occasion, she suffered the ignominy of being swept backwards while trying to negotiate a strong tide current, and the Gorleston crew soon returned the boat to Harwich. The persistent mechanical and technical problems from which City of Glasgow suffered eventually led to her replacement and withdrawal in May 1901, and then her sale out of service five months later for £100.

The third steam lifeboat: Queen

The RNLI's third steam lifeboat was named Queen to mark the sixtieth year of the reign of HRH Queen Victoria, and was launched in June 1897. The usefulness of a steam-powered lifeboat based on the Mersey had been proven by Duke of Northumberland, and so plans were formulated to build a new boat specifically for New Brighton. In planning a third steam lifeboat, the RNLI's committee discussed the issues widely and consulted the committee of the RNLI's Port of Liverpool Branch and George Watson, the Institution's naval architect, before going ahead with formulating the designs, no doubt taking into account the problems which had been experienced with the second steam boat.

One of the main problems in operating a steam lifeboat was, according to The Lifeboat, the 'great expense of upkeep', and a number of schemes were proposed to remedy this. The annual upkeep of a steam lifeboat

▼ The third steam lifeboat, Queen, during her trials on the Thames. A version of this photograph was published in The Lifeboat of 1 February 1898, but the industrial backdrop was removed, being substitued by a coastal landscape. (By courtesy of the RNLI)

THE THIRD STEAM LIFEBOAT QUEEN • TECHNICAL SPECIFICATIONS

DIMENSIONS • Queen was a little longer than either of her predecessors, being 55ft overall (53ft on the waterline), with a moulded breadth of 13ft 6in aft and an extreme breadth of 16ft; her moulded depth was 5ft 6in. Her draught on service was 2ft 9in forward, 3ft 4½in amidships, and 3ft 1½in aft; the greatest draught was amidships due to the intake through which the pump drew its propulsion water.

DISPLACEMENT • Fully equipped and ready for service, displacement was 31.87 tons.

MACHINERY • Queen was fitted with a near-vertical, 30-inch (762mm) diameter centrifugal pump, similar to that fitted in Duke of Northumberland, driven by a compound steam engine. This Thornycroft engine was supplied with steam by a Thornycroft patent watertube boiler whose working pressure was 145 pounds per square inch (10 bar). The boiler was fired by coal and oil, with an oil burner inserted into its furnace, but after four years, oil-firing was abandoned. The engine, with a stroke of 12 inches, had cylinders of 8½ inches and 14½ inches. Its exhaust steam was condensed in a seawater-cooled surface condenser.

POWER AND SPEED • With combined oil and coal, nine knots was obtained on the Mersey, and the indicated horse power was 222.2 at 400rpm; with coal only, 8.85 knots was achieved over the official double run on the Long Beach Mile, with an Indicated Horse Power of 206.5ihp at 403rpm.

[From information published in The Lifeboat, 1 February 1898, pp.1-6]

was approximately £800, the bulk of which comprised the salaries of the chief engineer, assistant engineer and two firemen, who were needed to maintain and operate the boat's machinery. So, when planning a new boat, it was decided to use a combination of oil and solid fuel, a technique perfected by James Holden, locomotive superintendent of the Great Eastern Railway, rather than just coal. This, it was hoped, would reduce the number of full-time crew needed from four to, at most, three.

The third steam lifeboat was a little longer than her predecessors, being 55ft overall, with a beam of 16ft, making her 5ft longer than Duke of Northumberland and 1ft 6in wider. Her displacement was 31.87 tons. Motive power was similar in many respects to the near-vertical centrifugal pump used in Duke of Northumberland, being driven by a compound engine supplied with steam by a Thornycroft patent water-tube boiler working at a pressure of 145 pounds per square inch (psi).

J.I. Thornycroft & Co, of Chiswick, undertook the construction of the new boat as well as its compound engine, with Greens sidelined perhaps because of the difficulties with City of Glasgow. The hull was made from steel and divided into numerous watertight compartments. The 200hp steam engine powered a centrifugal pump which was positioned with its shaft almost vertical, which gave the boat a top speed of about nine knots. Bunker capacity was three and a half tons of coal and two tons of oil. Comprehensive speed trials took place, with oil and coal both being used as the fuel on different occasions to assess which was most suitable. However, in the end it was found that the power to run the air compressor, driven from the main engine, which provided compressed air to atomize the boiler's oil fuel, reduced the power available to drive the boat. As a result, about four years into the boat's career, coal became the sole fuel, rendering the air compressor redundant.

The boat was completed in 1897 and evaluation trials were undertaken throughout the summer, with exhaustive tests undertaken on both the Thames and the Mersey. It was decided to complete the fuel consumption tests after the boat arrived at New Brighton and, on 8 October 1897, the boat left London for the Mersey. Basil Hall, Inspector of the Irish district, was in charge, and took the boat on an extensive passage up the east coast to Grangemouth, then through the Forth and Clyde Canal to Bowling, and from there down the west coast to the Mersey.

At times during the passage, the boat was running continuously for more than twenty-four hours. Coal was used, except for a short time between Grimsby and Berwick, when it was necessary to increase speed, and oil was tried, but not very successfully, for several hours. The boat seems to have behaved well and her engine was reliable, although no arduous weather was encountered. Considering the strain of such a long passage and that her machinery and boiler were new to the engineers, the RNLI deemed the voyage to have been 'highly satisfactory'.

The boat arrived at New Brighton on 27 October 1897, and the trials continued under the supervision of W.B. Cuming, the RNLI's consulting engineer. The boat had been partly funded by the RNLI's Port of Liverpool Branch, with a contribution from the Mersey Docks and Harbour Board (MDHB), the organisation responsible for the management of the Port, in accordance with their agreement with the RNLI. The remainder came from two legacies and the RNLI's general funds. The boat was named Queen, with the agreement of the Port of Liverpool Local Committee, to honour the Diamond Jubilee of Queen Victoria, Patron of the RNLI.

On 8 December 1897 a large-scale inauguration ceremony was held at Liverpool, during which the boat was formally handed over to the

▲ Queen on the River Mersey dressed overall, probably for her naming and inauguration ceremony in December 1897, shortly after she had taken up duty at New Brighton. (By courtesy of the RNLI)

RNLI and the Port of Liverpool Branch by Rear-Admiral Lord Charles Beresford, and christened Queen. Following the ceremony, the Lord Mayor of Liverpool, Lord Charles Beresford, Francis Henderson and several members of the Local Committee went aboard. Coxswain William Martin was at the helm, supervised by Basil Hall, and the lifeboat conducted 'a series of manoeuvres . . . which called forth the admiration of those on board and on shore'.

Although the steam boats at New Brighton were designated the No.2 lifeboats, they performed the majority of the rescues undertaken by the station during the first decades of the twentieth century. The first service Queen performed came on 25 February 1898, when she went out in heavy seas and a north-westerly gale to help the schooner Robert and Elizabeth, of Lancaster, which was aground off Crosby Beach. With huge waves repeatedly sweeping the schooner, Coxswain Martin skilfully manoeuvred the lifeboat close enough to rescue its crew of four, who were all completely exhausted. They were landed at New Brighton at midnight, just two hours after the lifeboat had launched.

STEAM LIFEBOATS

The impressive speed with which this rescue was effected could not have been equalled by the pulling and sailing lifeboats, even if they had received tug assistance.

The steam lifeboat proved to be ideal for rescue work in the Mersey and its various channels. The majority of the rescues involved assisting stranded steamers, such as on 7 March 1901 when Queen helped the steamship Dominion, of Liverpool, which was bound from Liverpool for Portland, Maine, with 270 persons on board. About ten minutes after the steam lifeboat arrived on scene, the Formby lifeboat John and Henrietta, a 35ft Liverpool pulling and sailing lifeboat, also arrived, and both lifeboats stood by the vessel until 8.30pm, at which point it floated off. The lifeboats then returned to their respective stations, with Queen towing the Formby boat as far as the Crosby lightvessel.

As had been the case with the RNLI's second steam lifeboat, Queen was also in need of continuous attention and repairs, and her full-time crew had their hands full maintaining her. Such work was not easy or straightforward, and safety procedures were often lacking. While Queen was undergoing repairs on 29 November 1905, two of her crew, Allan Dodd and John Jones, remained on board overnight, as night watchmen. It was bitterly cold and so they lit a fire in the stokehold to keep warm, then shut themselves in to try to maintain the warmth. Unfortunately, the next morning both were found dead, having suffocated during the night. The RNLI gave £225 to the dependents of the two lifeboatmen.

The boat performed her duties well when called on and another outstanding service was performed during the early hours of 27 August 1910 in a fierce westerly gale. The Mersey Docks and Harbour Board dredger Walter Glynn, with a crew of sixteen, capsized in violent seas off the North Wall. Some of the crew clung to the upturned dredger, while others attempted to swim ashore, with two losing their lives in the highly dangerous attempt. Queen was immediately called out under the command of Coxswain William Cross, finding several men clinging to the dredger's half-submerged superstructure.

The dredger was too close to the Wall for the lifeboat to be able to approach from leeward and so Coxswain Cross had to undertake a highly risky manoeuvre by coming in on the windward side. In the heavy seas, the difficult task of getting close to the wreck was made particularly

▲ The single-funnelled Queen on the Mersey with her crew on deck wearing cork life-jackets. She served New Brighton for more than a quarter of a century, saving almost 200 lives in that time. (By courtesy of the RNLI)

dangerous not only by the large amount of wreckage in the water but also by the dredger's funnels, her superstructure and dredge buckets, which all presented considerable hazards. But, after four attempts, Coxswain Cross succeeded in manoeuvring the lifeboat close enough so that the five men, who were clinging to the capsized boat, could be hauled to safety. One was seriously injured and getting him aboard the lifeboat without further injury involved considerable skill and teamwork by the lifeboat crew. Coxswain Cross then headed back to New Brighton, where the five survivors were landed. For his fine seamanship and considerable courage shown during this very difficult rescue, the RNLI accorded Coxswain Cross the Thanks of the Institution Inscribed on Vellum.

What proved to be the last service by Queen at New Brighton came in December 1923, when a new motor lifeboat was almost ready to take up duties at the station. Before being replaced, the old steam lifeboat helped to save 104 persons from the 12,269gt cargo steamer Armagh, of London, which had stranded in the Mersey in moderate weather. The lifeboat went out during the evening and immediately took off the only three passengers on board, leaving the crew, of 101, on board. The lifeboat then stood by while the crew tried to refloat the vessel, but, as was feared, the ship broke her back as the tide fell and a 15ft gap

▲ Queen during the later years of her career, by when she had been fitted with twin funnels. The auxiliary sails she carried, consisting of one lug and one jib sail, are raised. (By courtesy of the RNLI)

appeared in the deck. The lifeboat could not get alongside the forward part due to an insufficient depth of water, so a punt was obtained from the Dock Board's tender Vigilant, which was standing by. Manned by three lifeboat crew, this was used to transfer the steamer's crew to the lifeboat, in several trips. The rescued crew were later transferred to Vigilant, which brought them safely to the Liverpool Landing Stage.

Queen proved, during more than a quarter of a century at New Brighton, to be very successful. Between October 1897 and her withdrawal in August 1923 she launched on service eighty-one times and saved 196 lives. Of the six steam lifeboats, mechanically she was the best of the waterjet-propelled boats, although she still suffered with serious corrosion to her inlets. After being replaced at New Brighton, she was sold out of service at auction on 15 April 1924 for £75, with the RNLI receiving £54 3s 3d after deducting expenses, to Captain J.D.H. Filbee. He intended to use her for pleasure trips from New Brighton, but these were not viable so he sold her on to the shipping company Elder Dempster in 1925, who shipped her to Sekondi in West Africa. She later became a pilot tender in Lagos, Nigeria, working in this role for many years, until well after the end of the Second World War. What has become of her since is not known.

Screw-propelled steam lifeboats

After Queen, the third steam lifeboat, had entered service in late 1897, the RNLI considered building further such craft but reassessed the method of propulsion. The three boats built hitherto were all propelled by waterjets, so early in 1898 the RNLI's technical staff, in light of operational experience with steam plant and waterjets, considered the use of screw propellers. It was widely believed that a propeller would be vulnerable to damage in shallow waters or by fallen rigging and floating debris around stricken vessels, and the risk was deemed so great that it had hitherto mitigated against the use of propellers in lifeboats.

However, waterjet propulsion was also far from satisfactory. Although it had its benefits, a loss of speed was experienced when strong tides or currents were encountered, while considerable power was needed by

▼ The fourth steam lifeboat, James Stevens No.3, during her trials. The mast was subsequently moved aft of the funnels. This photograph appeared in The Lifeboat of 1 February 1899. (By courtesy of the RNLI)

STEAM LIFEBOATS

the pump to drive the boat, and even then a relatively low speed was achieved. Experience also showed that the pump was not immune from fouling, and on a few occasions the inlet passage in the boats' hulls had been clogged by seaweed, or the pump was damaged when stones, gravel and sand were drawn into it. On one occasion, one of the steam lifeboats was temporarily disabled after a rope was drawn through the inlet and wound round the spindle of the pump, bringing the engine to a stop. Clearing the pump could only be undertaken with the vessel out of the water, and this was a significant drawback.

As developments in screw propulsion were made towards the end of the nineteenth century, the RNLI looked to this method of propulsion as an alternative. A propeller in a protective cavity or tunnel in the hull had been used with success in gunboats on the Rivers Nile and Niger, and this suggested a way forward for using screw propellers in lifesaving work. The idea of protective tunnels convinced RNLI designers that propellers could work, and so two screw-propelled steam lifeboats were ordered from J. Samuel White, of Cowes. The order was placed on 14 January 1898 and the boats, funded from part of a large legacy left to the RNLI, were named James Stevens No.3 and James Stevens No.4.

Construction of the new boats was soon under way. By the autumn of 1898 James Stevens No.3 had been completed, and she became the RNLI's first screw-propelled lifeboat. She undertook her initial harbour trials at Cowes on 26 September 1898, with further trials being held on 11

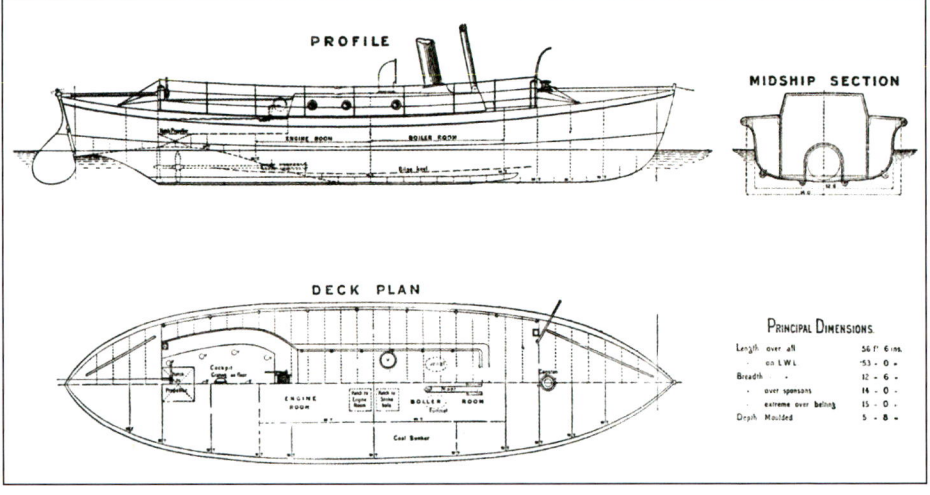

▼ Plans of the screw-propelled steam lifeboat, showing the mast forward of the funnel before modification, as in James Stevens No.3. This generic diagram appeared in The Lifeboat of 1 February 1899, page 273.

> ## JAMES STEVENS NO.3 and NO.4 • TECHNICAL SPECIFICATIONS
>
> **DIMENSIONS** • The two James Stevens steam lifeboats were 56ft 6in in length by 14ft in beam; their mean draught was 3ft 5½in over the bilge keels.
>
> **DISPLACEMENT** • Fully equipped, the displacement was 31.87 tons.
>
> **SPEED** • James Stevens No.3 reached a maximum speed of 9.59 knots on trials.
>
> **POWER** • The engine's mean indicated horse power on trials was 179.9, and mean revolutions per minute were 423.
>
> **MACHINERY** • The compound engine had a stroke of nine inches. Its high and low pressure cylinders had diameters of nine and sixteen inches respectively. The engine was supplied with steam by a White & Foster patent watertube boiler with a grate area of 15 square feet, a heating surface area of 500 square feet and a working pressure of 150 pounds per square inch.
>
> **PROPELLER** • The engine drove a four-bladed propeller, which was 3ft 1½in in diameter and 3ft 6½in In pitch.
>
> [From The Lifeboat, Vol.XVII, Feb.1899, pp.270-4]

October 1898, after which she was sent north to Grimsby, the station at the mouth of the Humber estuary for which she had been built, where she arrived on 19 October 1898.

James Stevens No.3 had an almost identical displacement to Queen, which was deemed the best of the three waterjet-propelled steam lifeboats, yet when Queen was compared with the new screw-powered boat she was found to be slower and consumed far more coal. James Stevens No.3 reached a maximum speed of 9.59 knots on trials, whereas Queen attained 8.83 knots, while the power necessary to drive Queen at a speed of just over eight knots was 168 indicated horse power (ihp), whereas the power required to obtain the same result from James Stevens No.3 was 97 ihp, or 61 ihp less than that of Queen, a reduction of almost sixty-three per cent in favour of the screw propeller. In terms of consumption, Queen burned 3 cwt 3 qtrs 18 lbs (199kg) of coal per hour to obtain eight knots, whereas James Stevens No.3 burned 2 cwt 1 qtr 23 lbs (125kg).

The two White-built steam lifeboats were more or less identical in design and layout. The only differences between them were a slight adjustment towards amidships in the position of the bunkers on the second boat, and the position of the mast. On James Stevens No.3 it was originally forward of the twin funnels, but was later moved aft, to prevent the lug sail from being burned. The mast on James Stevens No.4, as built, was aft of the funnels. As it was not always possible to obtain water for their boilers, the boats were each fitted with a Webster's evaporator, which were capable of providing three tons of distilled water from sea water in twenty-four hours. In compliance with what the RNLI saw as a necessary 'belt and braces' policy, which was not abandoned until all lifeboats had two engines and two screws, the steam lifeboats carried oars and sails in case the engine should fail.

James Stevens No.3 was intended to operate from Grimsby. Although Spurn Point, at the mouth of the Humber, was regarded as the ideal location for the lifeboat, Grimsby had better facilities for accommodation of crew members, firemen and supplies of coal, oil and water. The boat spent just over four years there, during which time was called upon relatively infrequently. She performed six services, of which only two resulted in effective rescues. She left Grimsby in January 1903 and next spent five years at Gorleston, undertaking thirty-five service calls in what was a rather more successful spell, being credited with saving thirty lives and assisting to save two steamships.

Eleven of these lives were saved from the 1892-built four-masted barque Optima, of Hamburg, on 19 January 1905. The barque stranded on the Haisborough Sands and guns fired by the light-vessels brought lifeboats and tugs to her assistance. She had been bound from Hamburg for Santa Rosalía, with coke, and had a crew of thirty-two. The steam lifeboat attempted to refloat her and the following day the Gorleston No.1 lifeboat and a private lifeboat were also called upon to help.

The following night gale-force winds endangered the barque's crew, who had to be taken aboard the lifeboats, but the next day operations were

▶ An early photograph of the 1898-built James Stevens No.3 taken during her trials. It is not known whether the people on board are crew or RNLI officials overseeing the trials.

▲ James Stevens No.3 at Cowes during an early demonstration run shortly after completion. Her mast was later moved abaft the funnels. (By courtesy of the RNLI)

resumed. By 21 January much of the cargo had been jettisoned and steam tugs arrived to tow her off. At 5pm the ship was bumping and straining, when suddenly her back broke and she swung round, putting everyone on board in considerable danger. The steam lifeboat, showing her superior power, managed to tow the other lifeboats clear, the Gorleston No.1 lifeboat suffering considerable damage as she struck the wreck before getting clear, while the barque eventually became a total wreck.

James Stevens No.3 was next sent to Angle, the Pembrokeshire station which covers Milford Haven, in November 1908. While there, she was called out a dozen times, but only on one occasion was she able to render any service. This was on 7 November 1911, when she was called out during a north-westerly gale to the aid of a boat from the ketch Trebiskin, of Padstow, which was adrift with one man on board. Also in distress were four men in a boat, which had put off from the small steamship Florence in an attempt to save the man from the ketch. Both boats were close to a dangerous shore against which they would might not have survived had they been wrecked. The lifeboat was able to steam close to each boat in turn and take them in tow, returning them to their respective vessels.

The most notable incident in which James Stevens No.3 was involved

STEAM LIFEBOATS

▶ The 1898-built James Stevens No.3 at Gorleston circa 1908, the second station she served. Her mast was originally forward of the funnels, but was moved abaft the funnel as the steadying sail tended to cause smoke to obstruct the view from the helm.

▼ James Stevens No.3 during her short spell as Angle lifeboat, when she operated from Milford Haven. (From an old postcard supplied by Iain Booth)

while she was at Angle came on 11 December 1914, when the lifeboat herself was in trouble after she broke her moorings in a severe gale and was washed ashore in Chapel Bay. The Haven was then on a war footing, so the War Department's vessel Haslar was tasked, and was quickly on the scene. With the help of Haslar's twelve-man crew, and seventeen men from the Royal Garrison Artillery and Royal Engineers, the lifeboat was quickly patched up and refloated. The master of Haslar, Mr Petty, was awarded a pair of binoculars and his men were given cash awards for

STEAM LIFEBOATS

▲ James Stevens No.3 alongside the Mail Pier, close to Admiralty Arch at Holyhead, with her eleven crew lining the deck wearing their kapok life-jackets. She spent five years at Holyhead until January 1928, at the end of her almost thirty-year RNLI career. (Supplied by Alf Pritchard)

their efforts, while the lifeboat was sent to Pembroke Dock for repairs. She had several new plates fitted around the hull, a new main keel and bilge keels and rudder repairs, which cost £414 7s 4d in total.

After being repaired, in August 1915 James Stevens No.3 was reallocated to Totland Bay, at the western end of the Isle of Wight, where she remained for the rest of the war. She had six calls, only one of which was effective, when she assisted the Belgian ketch Mercken on 11 February 1919. On 4 August 1919 she took part in the Thames Peace Pageant in London to celebrate the efforts of English mariners and merchant seamen during the First World War. After this, the most travelled of all the steam lifeboats next spent three years at Dover, reopening the station in October 1919; it had been closed in 1914 at the request of the naval authorities. She undertook five launches at Dover, but accomplished only a couple of services, in October 1920, standing by an Admiralty tug and saving a man from a fishing vessel.

Her final move was to Holyhead, where she replaced the veteran Duke of Northumberland in December 1922, and stayed until replaced by a motor lifeboat in December 1928. She ended her RNLI career with twenty service launches at Holyhead, being credited with saving eighteen lives and helping to save two vessels, as well as rendering assistance to six other craft. All of these were coasting schooners which, on the coasts of

◆ James Stevens No.3 at moorings at Admiralty Pier, Salt Island, Holyhead in the 1920s. (From an old postcard supplied by John Harrop)

the Irish Sea, only just remained economically viable for their owners as the steel motor ship started taking over. James Stevens No.3 was sold out of service in January 1929 for £125 by auctioneers in Holyhead. While in private ownership and in use as a pleasure boat, she was lost to the south-west of Porthdinllaen Point, off the Llŷn Peninsula, in July 1935.

The tragedy of James Stevens No.4

James Stevens No.4 was completed a few months after her sistership. She undertook her full speed trial on 29 December 1898 and passed her harbour trial at Southampton on 6 January 1899. She was sent to Padstow soon afterwards, arriving on 17 February, being designated the station's No.2 lifeboat. As described above, she was almost identical to her sister apart from the position of her bunkers and her mast. She was kept moored in the Camel Estuary ready to put out into the Celtic Sea, covering the often treacherous coast of North Cornwall. David Grubb, who had commanded Padstow's pulling lifeboat Arab since 1892, was appointed her Coxswain. Her crew was eleven in total, of whom the two engineers and two firemen were full-time employees of the RNLI and came to the station with the boat, while the rest were volunteers.

Within a few weeks of her arrival, on 7 April 1899, she was in action, heading north-eastward to the seas off Tintagel, where a vessel had been reported in distress. After searching for some time without result, the crew concluded that the alarm was false and they returned homeward. On the way back to station, however, they came across the French brig Emilie being towed by a steam tug, neither of which could make any headway. With the additional pulling power of the lifeboat, the brig, with a crew of seven, was towed to a safe anchorage in Hawker's Cove.

Sadly, James Stevens No.4 served at Padstow for little more than a year, before she was tragically wrecked in April 1900. The disaster that

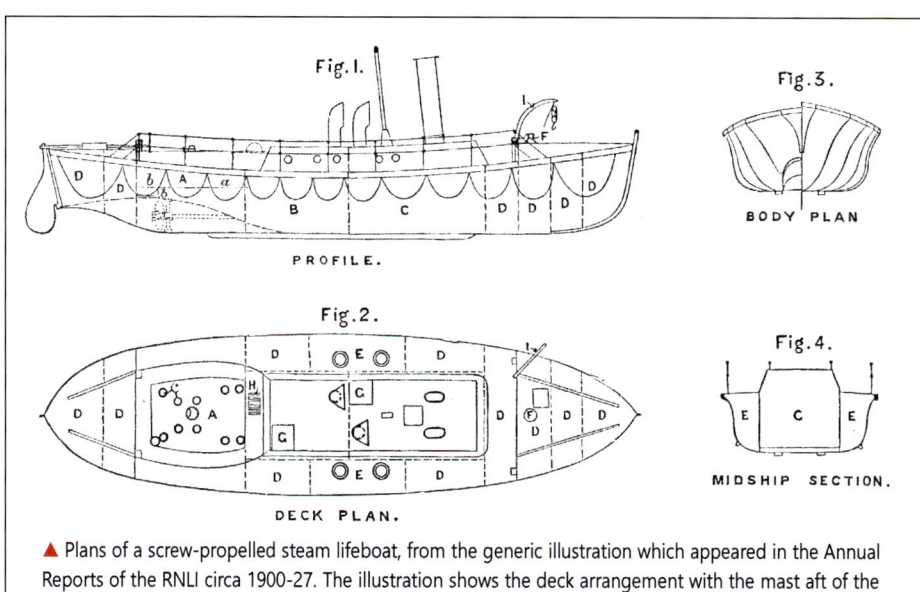

▲ Plans of a screw-propelled steam lifeboat, from the generic illustration which appeared in the Annual Reports of the RNLI circa 1900-27. The illustration shows the deck arrangement with the mast aft of the funnel, as in James Stevens No.4 and City of Glasgow (second), and James Stevens No.3 after alteration. The letters represent the following: (A) Cockpit (a deck, b propeller hatch, c relief valves), (B) Engine room, (C) Boiler room, (D) Watertight compartments, (E) Coal bunkers, (F) Capstan, (G) Hatches to engine and boiler rooms, (H) Cable reel, and (I) Anchor davit.

STEAM LIFEBOATS

befell her was one of the worst in the RNLI's history, for not only did the steam lifeboat capsize, but the station's pulling lifeboat, the 34ft self-righter Arab, on station since 1883, was also wrecked going to the same vessel, the ketch Peace and Plenty, of Lowestoft. The ketch, after spending the day of 11 April fishing, anchored in the lee of Stepper Point, but by nightfall began to drag her anchor in a strong westerly gale with very heavy breaking seas, and pilots put off to help. Before they could reach her, however, her anchor cable parted, and the vessel began to drift across the estuary into a dangerous bay. The Trebetherick Life Brigade fired several rocket lines to the casualty, one of which landed over the wreck, and five of the crew were dragged ashore, more dead than alive. Another jumped overboard and made it to the shore safely, but three others, were drowned. Meanwhile, the maroons were fired to summon the lifeboat crew, and both the steam and pulling boats put out.

Once she was at sea, Arab was taken around the Doom Bar, the dangerous entrance to the Camel Estuary, where the conditions were churning up huge seas. She was able to get towards the other side of the estuary via the narrow Britannia Way channel, thanks to her shallow draught. As she made her way to the casualty in the terrible weather, the ketch was being driven by the gale and huge seas further over the Doom Bar towards the rocks in Hell Bay and, in the darkness and blinding spray, the lifeboat crew could not see the casualty. In the gale force winds and driving surf they searched up and down, but could not make any headway towards Stepper Point, and were forced to stand by for over an hour in the heavy seas, waiting for a suitable opportunity to try to reach the casualty or attempt to get back ashore and to safety.

However, as the lifeboat was continuing to search, she was struck by a particularly heavy sea which unseated the crew, broke or carried away nine of the oars, and completely filled the boat. As the water drained away, the crew managed to regain the boat, albeit with great difficulty and only by grabbing the life-lines, and the anchor was immediately deployed in the hope that it would hold. A distress flare was lit to tell the steam lifeboat that Arab was in danger and needed help. As a return to Padstow was impossible, it was decided to run the boat ashore to the north east of the harbour, into a small creek, where it might be possible to beach her. So, using the spare oars, the crew let the anchor drag slowly,

◀ The remains of James Stevens No.4, barely recognisable as a steam lifeboat, after being thrown into a small cave at Greenaway Rocks with the loss of eight of her crew in the disaster of April 1900. One onlooker later commented she resembled 'nothing so much as a battered tin can'. (By courtesy of RNLI)

and, waiting for the lulls between waves, dropped gradually shorewards. They beached the boat and managed to scramble to the safety of the sand dunes, fortunately without any loss of life.

Meanwhile, James Stevens No.4 headed out beyond the estuary mouth and made for deeper water, as her draught was too great for a direct crossing of the shallow estuary. Coxswain Grubb had seen the lifeboat's flare, but believed it came from the ketch, so steamed out to deep water, before turning shorewards. However, before completing the turn, James Stevens No.4 was caught by a huge sea, which lifted her stern completely out of the water, spun her broadside to the waves, and turned her over. Although the seven crew in the cockpit were thrown

REMEMBERING THOSE WHO GAVE THEIR LIVES IN 1901

MACHINERY • Those trapped in the machinery spaces on board James Stevens No.4 were Chief Engineer John Martin, Second Engineer James Old and the two firemen, Joseph Stephens and Sydney East.

OTHER CREW • The other Padstow crew who lost their lives were Coxswain David Grubb and crew members Edward Kane, John Bate and James Grubb.

MEMORIAL • A number of memorials were erected in memory of the lost crew of James Stevens No.4, including an impressive one in Padstow cemetery, funded by public subscription

STEAM LIFEBOATS

clear, the men in the boiler room and engine room were trapped. Three of the men washed off the boat were swept ashore and revived, but tragically four others did not survive. The lifeboat was thrown into a small cave at Greenaway Rocks a complete wreck.

When news of the disaster spread to the county and then nationwide, sincere sympathy was expressed for the relatives of the deceased, and a relief fund was opened for the dependents of the eight men, who left behind four widows and fourteen young children. The RNLI approved a £1,000 contribution to the fund and also made extra payments to each of the crew of Arab, as well as the deckhands on the steam lifeboat. HM Queen Victoria contributed £25 to the fund and HRH The Prince of Wales £20. The medical expenses of the survivors and funeral expenses of the four men whose bodies had been recovered were defrayed by the RNLI. The Board of Trade enquired into the cause of the disaster, but could not attribute any fault to the lifeboat or mismanagement by the crew, and the verdict on the victims was 'death by misadventure'.

The last RNLI steam lifeboat

As the first City of Glasgow had suffered from so many reliability and mechanical problems while she was at Harwich, a new steam lifeboat was ordered in 1900 for the station, which was considered important enough to warrant a steam lifeboat, whose power was deemed essential in order

▼ The second City of Glasgow in the Pound at Harwich, with her engineers aboard, and the steam tug Merrimac moored beyond her. (Supplied by Ken Brand)

▲ The 1901-built City of Glasgow shows an impressive turn of speed. Power came from a single vertical compound engine of 180hp running at 410rpm. (By courtesy of Harwich RNLI)

to reach the outlying sandbanks where most incidents occurred. A boat identical to James Stevens No.3 and No.4 was ordered from J. Samuel White at Cowes. The new boat was named City of Glasgow, like her predecessor, in recognition of the gift of the first boat by the citizens of Glasgow, following Lifeboat Saturday demonstrations of 1893 and 1894.

The second City of Glasgow was driven by a propeller, which was recessed in a tunnel so she could 'proceed in shallow water or rocks in perfect safety'. The boat and her machinery were put through a series of trials off Cowes, Isle of Wight in May 1901, which were deemed successful, after which she left for Harwich. The trials showed that, compared with the other steam lifeboats, the second City of Glasgow was the fastest. She achieved an average maximum speed on the measured mile of 9.66 knots per hour, with an ordinary working full speed of 9.143 knots.

City of Glasgow left her builder's yard in early May 1901 and headed east, crossing the Thames estuary for Harwich, where she was placed on station on 9 May. During the morning of 13 May, she was taken on a trial trip to Walton in a stiff breeze and choppy seas, covering nine miles in fifty minutes on an ebb tide, and, according to the local newspaper

STEAM LIFEBOATS

report, 'the manner in which the boat behaved was extremely gratifying to the crew'.

City of Glasgow completed ninety-nine services, saved eighty-seven lives and either saved or assisted to save ten vessels during her time at Harwich. One of her most outstanding rescues was on 7 April 1908 when she went to the French schooner Notre Dame de Toutes Aides, which was stranded on the Kentish Knock sandbank. The Margate lifeboat had launched but was unable to make progress in the extremely heavy seas and had to return to harbour. HMS Dreadnought was in the vicinity and had launched a cutter, but after some time fighting the conditions, this had to return to its parent ship. The steam lifeboat put out and, once on scene, went alongside the Frenchman, but the seas caused her to range so much that twice heavy hawsers were snapped.

The Coxswain then steamed head-on to the vessel, bringing the lifeboat as close as he dared. He got close enough for the schooner's crew to jump into the lifeboat, but it took five approaches to take off all of the nine men on board. One of the schooner's men had been washed away earlier and another, who tried to get away in the schooner's own boat, was also drowned. The rescued men were extremely cold from exposure, but they were revived with hot coffee and biscuits. As City of Glasgow left the scene, the Ramsgate lifeboat arrived under tow from a steam tug, but there was no chance of saving the vessel.

▶ The second City of Glasgow on service at Harwich to the schooner Notre Dame de Toutes Aides, of Nantes, on 7 April 1908, as depicted in The Lifeboat journal. (By courtesy of the RNLI)

▲ The second City of Glasgow moored in the Pound at Harwich with the 43ft Watson sailing lifeboat Ann Fawcett beside her. Ann Fawcett was built in 1904 and spent just over eight years at Harwich working with the steam lifeboats, after which she was transferred to Dun Laoghaire. (By courtesy of Ken Brand)

Afterwards the French captain wrote a touching letter to the local Press thanking everybody for their help. He mentioned that the lifeboat crew took off their own clothes to cover the rescued for the long journey back to Harwich in the open cockpit of the lifeboat, and the RNLI concluded the official account by stating: 'The service of the Harwich boat was a good one and splendidly performed, dogged determination playing a conspicuous part in it.' Subsequently, at the International Conference on Life-Saving held at St Nazaire and Nantes in August 1908, diplomas were conferred upon the Coxswain, crew and engineers, and each diploma was accompanied by the silver medal of the La Société des Hospitaliers Sauveteurs Bretons.

Another fine service took place on 23 September 1915, when the 9,181gt Dutch liner Koningin Emma, homeward bound from Java, fouled a mine and went aground off Walton-on-the-Naze, a mile west of the Sunk lightvessel. She had about 150 passengers on board and a crew of fifty, and was bound for Amsterdam. Government vessels had taken off the greater part of these when the lifeboat arrived. The steamship had a severe list and her master asked the Coxswain to stand by until daylight when they could assess the situation. When the tide started to make, the steamship began to fill and heel over and the Captain and

STEAM LIFEBOATS

◀ An unusual photo of the second City of Glasgow, with, on the port side, George Armstrong (chief engineer). It is not known where or when this photo was taken. She is on the wheeled carriage of a patent slipway, used to haul small vessels out of the water as a cheaper alternative to drydocking, to facilitate maintenance that is not possible when the boat is afloat. (From an old postcard supplied by Ken Brand)

his remaining crew, twenty all told, scrambled into the lifeboat. The lifeboat crew steamed around the vessel and it was obvious she would become a total wreck, so most of the crew were taken off and put aboard Batavia, another steamer, and the lifeboat saved a further twenty.

The final service undertaken by City of Glasgow was another long one. She launched at 4.45pm on 7 January 1916 to the steamship Zeeland, of Rotterdam, which was aground near the Longsand Lightship in gale force winds and very heavy seas. The lifeboat stood by all night and in the early morning, together with the tug Thames, assisted in refloating the vessel, which was able to resume its voyage unaided.

In October 1917 the Admiralty wrote to the RNLI requesting the use of City of Glasgow, stating that, while they were not seeking to take over many lifeboats, the steam lifeboat was an exception. The RNLI suggested they purchase the boat, and so in December 1917 the Admiralty paid £4,290 for the boat and her equipment. Following this, the Harwich station was closed and the Coxswain, Adam Garnett, was pensioned off. Fireman Charles Knight was given a gratuity and pension, while Fireman Donald Wood was transferred to the steam lifeboat at Dover, James Stevens No.3.

The RNLI's unique steam tug

Following the tragedy at Padstow in April 1900, the RNLI considered replacements for the two lifeboats that had been lost and took an unusual step, as The Lifeboat of May 1901 explained: 'Having regard to the terrible nature of the disaster and the exceptional requirements of the locality, the Committee have decided not to replace the wrecked Steam Life-boat, but to substitute for it a sailing Life-boat to be towed by a Steam-tug which has been specially designed for the purpose by Mr G.L. Watson, the Consulting Naval Architect of the Institution and which is now being built.' This move resulted in the creation of a unique arrangement at Padstow with the operation of the only steam tug the RNLI ever built. The RNLI had no option but to order its own tug in what was described as 'an entirely new departure' for the Institution, because no commercial tug was available at Padstow.

The steam tug arrived at Padstow in early September 1901. She was classed 100A1 in Lloyd's Register of Shipping, measured 95ft 6in by 19ft and was built of steel. A Scotch boiler and two direct-acting, inverted

◀ A fine photograph of the unique steam tug Helen Peele at moorings in the Camel estuary at Padstow. She was primarily intended for operations in conjunction with the sailing lifeboat Edmund Harvey, which was designated the station's No.2 lifeboat.

STEAM LIFEBOATS   **67**

▶ The unique steam tug was built for Padstow as it was considered something even heavier than a steam lifeboat was needed to cope with the severe conditions at the entrance to the River Camel. This picture gives an idea of the extent of the surf enountered in the estuary during westerly gales. On this occasion, 1 September 1908, the steam tug was towing out the lifeboat Edmund Harvey to escort the ship Talus, of Greenock, to safety around Hartland Point. (By courtesy of the RNLI)

compound condensing marine engines, indicating collectively 300hp with natural draught, and 400hp collectively with mild forced draught, drove twin screws. During her trials, she reached a speed of over ten knots, which was regarded as 'very satisfactory'. Built by Ramage and Ferguson Ltd, of Leith, she was launched from their yard on 28 June 1901, being formally named at the yard by Miss Ramage. Trials took place between 2 and 5 September 1901, when the twin-screw tug was found to be very manoeuvrable when going both ahead and astern.

A few days later the tug left Leith and headed south, calling at Southend-on-Sea to collect the new 42ft twelve-oared sailing lifeboat built by Thames Ironworks, which she towed to Padstow. The tug was named Helen Peele, and her cost was defrayed in part from a special bequest by the late C.J. Peele, of Chertsey. The pulling and sailing lifeboat was named Edmund Harvey after the donor who gave the RNLI a generous £3,500 legacy. The two new boats arrived at Padstow on 15 September 1901, but had a somewhat ignominious arrival as the tug grounded on the Doom Bar as she entered the harbour, fortunately floating off about three hours later without being damaged.

When Helen Peele arrived, a second spare propeller was provided so she carried both port and starboard propellers. A medicine chest was supplied at a cost of £5 5s, with the tug's permanent crew attending St

John Ambulance classes at Padstow so they knew how to use its contents. This was one of the earliest examples of the RNLI providing first aid training. A number of men were employed full-time to maintain, operate and crew the steam tug: the permanent crew consisted of master, mate, cook, two engineers and a fireman, and the engine room was staffed by an engineer, an assistant engineer and a fireman. Henry Richards, aged thirty-one, was appointed the tug's chief engineer on 17 June 1901.

The crew had to have various qualifications and skills, with the master needing a certificate recognised by the Board of Trade, experience in the management of a steam tug and to be 'thoroughly acquainted with the Cornish coast and Bristol Channel'. In the engine room, the RNLI specified that neither of the engineers should be more than thirty-five years of age. The master's salay was not to exceed £2 5s a week, the mate's not to exceed £1 15s, the engineer's not to exceed £2 5s, and the second engineer's was not to exceed £1 15s. The 'Forecastle Cook', as the RNLI termed his position, had to be a qualified mariner, no older than thirty-five, with a salary of 28s per week. In addition to the permanent crew, six volunteer deck hands were taken when the tug went out.

Both the steam tug and the large lifeboat were kept at moorings off Hawker's Cove in the Camel Estuary, although initially the tug had to be secured alongside the quay or anchored in the pool off the harbour as the moorings in the estuary were not ready. The two craft successfully undertook a considerable amount of life-saving work during the twenty-

▼ Helen Peele moored alongside the docks at Bristol during a visit to the city. (By courtesy of Padstow RNLI)

eight years they were stationed at Padstow, and Helen Peele was credited with twenty-four service launches and twenty lives saved on her own. Edmund Harvey launched forty-one times and saved seventy-eight lives.

The first service performed by Edmund Harvey and Helen Peele working together took place on 14 December 1901 and proved to be exceptionally testing, especially for the men being towed. The rescue craft put out at 6.30pm into a strong north-easterly gale, rough seas and squally weather. Even in the estuary there were huge waves and the first tow rope broke, forcing Edmund Harvey's crew to anchor while the tug reconnected the line. When they eventually got to sea, the crew of Edmund Harvey were already soaked to the skin, but they continued westwards, drenched in spray, and found the steamship Auguste Legembre, of Algiers, which was on passage from Barrow to Port Talbot, with about thirty persons on board. Her steering gear had broken, and, when the lifeboat and tug reached her, she was sinking.

A commercial tug took the steamer in tow, but the rope parted and fouled the tug's propeller. Reconnecting the tow in the darkness was out of the question so Helen Peele faced the gale through the night, steaming ahead just sufficiently to keep her station, with Edmund Harvey hanging on astern. The tug's crew could at least make themselves meals and hot drinks, but the position of the fifteen men in the lifeboat was dreadful. Their emergency rations consisted of chocolate, biscuits and brandy, while they huddled under the sails seeking shelter from the wind and spray. The next morning two more tugs arrived and all four commenced to tow the disabled steamer, with the lifeboat stationed astern.

Conditions had improved, and the slower pace of the tow stopped the waves coming over the bows so that the lifeboat crew's clothes began to dry. Later in the day, as they rounded Hartland Point, the sky became overcast and snow fell, with 'their eyes almost blinded with salt, throats sore and dry, and their hands and feet swollen almost beyond recognition', according to historian Claude Berry's account. Eventually they arrived at Cardiff at noon on 16 December, having spent forty-four hours at sea, frozen and soaked most of the time, but all survived the ordeal. The lifeboat was towed back to Padstow the next day by the steam tug.

During 1917 Helen Peele was requisitioned by the Admiralty, being commissioned as an RN tug at Swansea on 29 August 1917. She

▲ Helen Peele pictured at Cowes. The steel-hulled vessel had a fine career with the RNLI, and remains the largest vessel employed by the RNLI for rescue work in the British Isles. (By courtesy of the RNLI)

proceeded to Portland, via Penzance, and, after being refitted, carried out target towing duties for HMS Amphitrite. During November 1917 she assisted in refloating the steamship Alice Taylor in Weymouth Bay and later towed the steamer Briez into Portland. Helen Peele went to South Shields in December 1917, before going further north to Lerwick the same month. In February 1918 she was refitted at Aberdeen, returning to Lerwick on 16 March, after which she was used to assist small ships in distress. She remained at Lerwick until early 1919, when she was released from RN service. In February 1919 she was repaired and repainted at the Admiralty's expense at Harris Brothers boatyard in Swansea, where the surveyor found the hull and machinery to be in generally good condition, and by early May 1919, with repairs complete, the tug had been handed back to the RNLI.

During the 1920s Helen Peele continued in service at Padstow, although between 1921 and 1928 neither she nor Edmund Harvey were called upon. However, what proved to be the last service undertaken by Helen Peele saw the tug involved in a fine medal-winning service, enabling her to leave the RNLI fleet in something of a blaze of glory. In the early hours of 27 November, one of the local fishing fleet, Our Girlie, of Port Isaac, with five men on board, was reported overdue in a

▶ This fine photograph of the steam tug Helen Peele moored in the harbour at Padstow gives a good impression of her deck layout. (By courtesy of Padstow RNLI)

north-westerly gale and very heavy seas. It was still dark, but fortunately the tug's searchlight picked out the missing boat anchored close to a lee shore at Port Quin, in great danger of being wrecked on the rocks. There was no time to call for one of the lifeboats, so Captain J. Atkinson decided to make a hazardous attempt to reach the casualty using the large tug. He ordered oil be discharged to calm the seas, and then went to within 200 yards of the rocks, and anchored. The water was shallow, but he had the cable run out and, despite the considerable risk, successfully got the tug alongside so that the fishing boat's crew could climb to safety. As the windlass drew in the chain pulling the tug back towards deeper water, the fishing boat's cable parted and she was wrecked. For this outstanding service, Captain Atkinson was awarded the RNLI's Bronze Medal, the Thanks of the Institution Inscribed on Vellum were accorded to each of the crew, and to Captain E.P. Hutchings, the Honorary Secretary, who was also presented with an inscribed barometer.

At the end of April 1929 a powerful new twin-screw Barnett motor lifeboat was sent to Padstow to replace both the steam tug and sailing lifeboat. Helen Peele left Padstow for the Clyde on 2 May 1929, having been sold out of service to a Captain John Turner to become a yacht tender; what became of her after this is not known.

Careers of the Steam lifeboats

The six steam lifeboats and one steam tug operated from stations where they could be kept afloat. At the time, suitable harbours were fewer in number than they are now. Therefore, places like Grimsby, Milford Haven and Totland Bay were deemed suitable, as well as Holyhead, Harwich and New Brighton (the latter covering Liverpool's approaches). All of these were regarded as ideal stations for steam lifeboat operation. Indeed, Harwich was served by three different steam lifeboats, while two served at both Holyhead and New Brighton. The last steam lifeboat in service was the much-travelled James Stevens No.3, while the last steam-powered vessel in RNLI service was the steam tug Helen Peele, which was withdrawn in 1929.

▼ The 1901-built City of Glasgow (ON.446) on station at Harwich. Her RNLI service lasted from 1901 to 1917. (By courtesy of Ken Brand)

Duke of Northumberland

Key data
BUILDER 1889, R. & H. Green, Blackwall; yard no.G 227
DONOR RNLI general funds
DIMENSIONS 50ft x 14ft 3in x 5ft 9in
STATIONS Harwich Sep 1890 – Jul 1892, Holyhead Oct 1892 – May 1893, New Brighton May 1893 – Dec 1897, Holyhead Dec 1897 – Nov 1922
RECORD 15 launches, 33 lives saved (Harwich); 6 launches, 9 lives saved (Holyhead); 29 launches, 14 lives saved (New Brighton); 131 launches, 248 lives saved (Holyhead)
DISPOSAL Sold out of service in January 1923 for £70

Service career
DUKE OF NORTHUMBERLAND served at Harwich for less than two years, undertaking a first service on 8 October 1890, within a month of arriving on station, when she stood by the brigantine Ada, of Faversham, which was aground on the Cork Sand. At high water the brigantine was towed off by a steam tug and taken to Harwich. In 1892 Duke of Northumberland was transferred to Holyhead, where she served until May 1893. She was then sent to New Brighton, arriving on the Mersey on 17 May 1893. She was initially to stay there for a year of trials, but ended up spending more than four years at the Mersey station, launching twenty-nine times and saving fourteen lives.

Her first effective service at New Brighton came on 27 January 1894, after a vessel had been reported to be aground on Taylor's Bank at the entrance to the Mersey. The Duke slipped her moorings with Coxswain William Martin at the helm, and headed out into heavy seas and a south-westerly gale. The lifeboatmen found the schooner Maria Lamb, of Runcorn, and her crew of six, along with their dog, were rescued and landed at New Brighton. The steam lifeboat was ideal for work in the Mersey estuary, which is why another steam lifeboat, Queen, was built for the station. After being replaced on the Mersey by Queen, Duke of Northumberland went back to Holyhead, where she went on to enjoy a glorious career, which included the Gold Medal service to the steamship Harold, as described in the main body of the text.

The last services performed by Duke of Northumberland were on 15 January 1922 to the Royal Mail steamship Hibernia, which was in difficulty 200 yards from the breakwater in a strong gale. Her final service was undertaken on 21 October 1922, and involved standing by the steamship Jolly Helen in rough seas, half a mile north east of the Skerries.

ON.231

After service

DUKE OF NORTHUMBERLAND was sold out of service at auction in January 1923 for £70, and was used a workboat on the Mersey. However, nothing more is known of her subsequent use or location until the 1990s. Then, in 1994 she was found in a completely derelict condition at the old Ready-Mix Concrete works, Terrace Road, Widnes, where she was slowly being covered in waste concrete. The works were subsequently demolished and the land was redeveloped, but some remains of the boat can still be found on the river bank. A sad end to a revolutionary lifeboat.

▲ Duke of Northumberland on station at New Brighton, with her crew wearing cork life-jackets.

▲ The remains of Duke of Northumberland in the river bank at Widnes, pictured in 2013. (Martin Fish)

STEAM LIFEBOATS

City of Glasgow

Key data
BUILDER 1894, R. & H. Green, Blackwall; yard no.G 289
DONOR City of Glasgow Lifeboat Fund
COST £2,639 10s 0d
NAMED 16 June 1894 at Glasgow by Mrs Bell, wife of the Lord Provost of Glasgow
DIMENSIONS 53ft x 16ft x 5ft 6in, 48ft waterline
DISPLACEMENT 28 tons
STATIONS Harwich No.2 Nov 1894 – Nov 1897, Gorleston Nov 1897 – Feb 1898, Harwich Feb 1898 – Oct 1901
RECORD 9 launches, 4 lives saved (Harwich); 1 launch, 0 lives saved (Gorleston); 5 launches, 28 lives saved (Harwich, second term)
DISPOSAL Sold out of service in October 1901 for £105 to Mr M. Lemon, Harwich

Service at Harwich and Gorleston

CITY OF GLASGOW, the second steam lifeboat, was not particularly successful and suffered from reliability issues and mechanical problems, her operational career lasting less than a decade. Her first effective service at Harwich was completed on 6 June 1895, when she went out in rough seas to the schooner Hans, of Rendsburg, which was leaking badly and was aground on the West Rocks. Some of the lifeboatmen boarded her, and she was taken in tow. City of Glasgow's service at Gorleston lasted just three months, with the crew there not impressed when she was swept backwards by a strong current.

The final service undertaken by City of Glasgow came on the morning of 30 March 1901 and saw her crew complete a fine service, to the schooner Rose, of Ipswich, in a severe gale. The strong winds and heavy seas made approaching the wreck difficult but, after several attempts, and with a crowd of onlookers cheering them on, the lifeboat crew succeeded in getting a line aboard, and then getting the lifeboat close enough for the sailors to jump to safety. Three men were rescued, but the fourth slipped between the lifeboat and the casualty. Fortunately, one of the lifeboat men caught hold of him and hauled him on board the lifeboat, although he was injured.

A Mr J. Flory, of the Missions to Seamen, was determined that the crew should receive formal recognition for their work, and, at the Grand Nautical Fair at Ipswich on 30 April 1901, with the Mayor overseeing proceedings, the Mayoress presented Silver Medals to the twelve lifeboatmen involved. The name of each was inscribed in capital letters

ON.362

on the medal, together with the words: 'Memento for gallantry in saving the crew of the schooner Rose, of Ipswich, 30th March 1901. Let not the deep swallow me up'. The recipients were: Coxswains Benjamin Dale and William Tyrrell, James Dale, Matt Scarlett, Ben Dale, jnr, A. Garnett, John Garnett, George Fenner, R. B. Scott, P. Harley, James Melville, and Charles Knights. The medals were unconnected with the RNLI, but were funded and presented through the efforts of local people.

After service

CITY OF GLASGOW was sold out of service in October 1901, but nothing is known of her subsequent career, although she possibly remained at Harwich initially.

▲ The first City of Glasgow on trials, probably in the Thames Estuary. She proved to be less than satisfactory and her career at Harwich was blighted by technical problems. (By courtesy of the RNLI)

STEAM LIFEBOATS

Queen

Key data
BUILDER 1897, J.I. Thornycroft & Co Ltd, Chiswick yard no.TH325
DONOR Two legacies, RNLI Funds and the Port of Liverpool Branch of the RNLI
COST £5,145
NAMED 8 December 1897 at Liverpool
DIMENSIONS 55ft x 16ft 6in x 5ft 6in
STATIONS New Brighton Oct 1897 – Aug 1923
RECORD 81 launches, 196 lives saved
DISPOSAL Sold out of service in April 1924 at auction for £75

Service at New Brighton
QUEEN was the second steam lifeboat stationed at New Brighton, and she arrived on the Mersey on 27 October 1897. She continued her trials under the supervision of W.B. Cuming, the RNLI's consulting engineer, once she was at her new station. Funding for the boat had been raised locally, with £1,022 14s 3d coming from the RNLI's Port of Liverpool Branch, and £1,000 contributed by the Mersey Dock and Harbour Board. The remainder came from two legacies amounting to £1,474 17s 0d, and the RNLI's general funds. The boat was named Queen, with the agreement of the Port of Liverpool Local Committee, to honour the Diamond Jubilee of Queen Victoria, Patron of the RNLI. On 8 December 1897 an inauguration ceremony was held at Liverpool for the new boat, during which she was formally handed over to the RNLI and christened Queen. She enjoyed an illustrious career on the Mersey, undertaking many fine rescues and being credited with saving 196 lives.

After service
QUEEN was sold out of service in April 1924 at auction for £75 to Captain J.D.H. Filbee, who intended to use her for pleasure cruises on the Mersey from New Brighton. When the pier tolls were not reduced for his planned landings, he decided that his venture would not be viable so he sold her to the shipping company Elder Dempster in 1925. They altered her before shipping her to Sekondi, in the Gold Coast (now Ghana), for use in ferrying passengers between their ships and the shore. She was taken to Sekondi, where it was intended that she would carry passengers to and from the shore. How long she remained in this role is not clear as a deepwater port at Takoradi was built to the west, effectively replacing Sekondi, and Queen may have then moved there. She was later reported to be in use as a pilot boat at Lagos, Nigeria.

ON.404

▲ Queen undergoing trials, probably around the Thames Estuary. The boat underwent a 'consumption trial' on 25 August 1897, to test measure her fuel consumption when burning coal alone or coal plus oil. Each trial lasted two hours, and on board at the time were Chief Inspectors of the French and German lifeboat societies.

▲ Queen being secured on deck aboard Elder Dempster's steamer Egori in Liverpool Docks in 1926 to be taken to West Africa to be used to as a tender between ships and the shore; note the wooden benches and high rails installed for her new role. (By courtesy of the RNLI)

STEAM LIFEBOATS

James Stevens No.3

Key data
BUILDER 1898, J. Samuel White, Cowes, yard no. W 1054
DONOR Legacy of James Stevens, Birmingham
COST £3,298
DIMENSIONS 56ft 6in x 15ft 9in x 5ft 8in
STATIONS Grimsby Oct 1898 – Jan 1903, Gorleston Jan 1903 – Nov 1908, Angle Nov 1908 – Aug 1915, Totland Bay Aug 1915 – Aug 1919, Dover Oct 1919 – Dec 1922, Holyhead No.1 Dec 1922 – Jan 1928
RECORD 6 launches, 0 lives saved (Grimsby); 37 launches, 30 lives saved (Gorleston); 12 launches, 5 lives saved (Angle); 6 launches, 0 lives saved (Totland Bay); 5 launches, 1 lives saved (Dover); 20 launches, 18 lives saved (Holyhead)
DISPOSAL Sold out of service in January 1929 for £135 by Preston Thomas & Co. Auctioneers at Holyhead

Service career
JAMES STEVENS NO.3 was much travelled in her operational career with the RNLI, serving no fewer than six different stations and being the only steam lifeboat to operate from Angle, Totland Bay (on the Isle of Wight) or Dover. She was built for Grimsby, to cover the busy Humber Estuary, but was not particularly successful there, so moved to Gorleston. Her longest period of service came at Angle, in west Wales, where she spent almost seven years. She spent most of the First World War at Totland Bay, on the western tip of the Isle of Wight, where she was kept moored off the slipway used by the pulling and sailing self-righting lifeboats. On 4 August 1920 she took part in the water pageant on the Thames to celebrate the Allied victory, by which time she was in service at Dover. Her final move was to Holyhead in Anglesey, where she remained until December 1928. Her last service at Holyhead, on 10 February 1928, proved to be the last service by an RNLI steam lifeboat. She rescued the four crew from the schooner Agnes Glover, of Castletown, which was in distress off Penrhos Point, in heavy seas and a south-westerly gale.

After service
JAMES STEVENS NO.3 was sold out of service in January 1929 and became a pleasure boat, being renamed Helga. While in private ownership, she was lost when about some two and a half miles south-west of Porthdinllaen Point, on the Llyn Peninsula, at the beginning of July 1935.

ON.420

▲ James Stevens No.3 at Dover in the years after the First World War; during a thirty-year career, she became the most travelled of any RNLI steam lifeboat. (By courtesy of the RNLI)

▲ James Stevens No.3 in private hands as the pleasure boat Helga.

STEAM LIFEBOATS

James Stevens No.4

Key data
BUILDER 1899, J. Samuel White, Cowes, yard no. W 1055
DONOR Legacy of James Stevens, Birmingham
COST £3,339 13s 4d
DIMENSIONS 56ft 6in x 15ft 9in x 5ft 8in
STATIONS Padstow No.2 Feb 1899 – Apr 1900
RECORD 4 launches, 9 lives saved
DISPOSAL Wrecked on service 11 April 1900

Service at Padstow
JAMES STEVENS NO.4 served at Padstow for less than a year. She was completed in December 1898 a few months after her sistership, passed her harbour trial at Southampton on 6 January 1899 and was accepted by the RNLI on 18 January 1899. With four tons of coal in her bunkers and boiler room, as well as full fresh water tanks, her complete equipment and 15cwt of deadweight in the cockpit to represent survivors and nine crew, her draught was 3.1ft forward, 3.9ft aft and a mean of 3.5ft. She had a loaded displacement of thirty-one tons, was fitted with nine relieving tubes, and her engines gave a mean indicated horse power of 202.075 during trials, operating at a mean 443.78 revolutions per minute, giving her a mean speed of 9.339 knots on the measured mile. She carried a lugsail and a foresail.

Career-ending tragedy
DURING her short career at Padstow, James Stevens No.4 saved nine lives, all of them during 1899, and launched four times on service, three of which were undertaken during the year she arrived. The disaster that ended her career claimed the lives of eight of her crew. Following the capsize, James Stevens No.4 was totally wrecked, and was barely recognisable, after being thrown into a small cave at Greenaway Rocks with the loss of eight of her crew in the disaster of April 1899. The seven crewmen who were in the cockpit were thrown clear, but the four men tending the boiler and in the engine room were trapped. Three of the men washed off the boat were swept ashore and revived, but tragically the other four did not survive. The lifeboat was thrown into a small cave in the rocks at Hell Bay and was a complete wreck resembling, one onlooker commented the following day, 'nothing so much as a battered tin can'.

ON.421

▲ The steam lifeboat James Stevens No.4 was twin funnelled; she was propelled by twin screws. (By courtesy of Padstow RNLI)

▲ James Stevens No.4 passing Stepper Point on her way to the ketch Peace and Plenty in April 1900, shortly before she was tragically wrecked. (By courtesy of Padstow RNLI)

City of Glasgow

Key data
BUILDER 1901, J. Samuel White, Cowes, yard no. W 1101
DONOR City of Glasgow Lifeboat Fund
COST £4,191
DIMENSIONS 56ft 6in x 15ft 9in x 5ft 8in
DISPLACEMENT 32.6 tons
STATIONS Harwich May 1901 – May 1917
RECORD 99 launches, 87 lives saved
DISPOSAL Sold December 1917 to Admiralty, and renamed Patrick

Service at Harwich
CITY OF GLASGOW arrived at Harwich in May 1901 direct from her builder's at Cowes. Her steel hull included a 12ft gunwale, was fitted with ten relieving tubes, had iron side keels 34ft long and bilge keels 25ft long. She was reputedly the fastest of the steam lifeboats, being capable of a top speed of 9.66 knots. She spent sixteen years at the important station covering the northern reaches of the Thames Estuary, and working in conjunction with a succession of large sailing lifeboats for much of that time. In October 1917 City of Glasgow was taken over by the Admiralty and the Harwich station was closed, not being reopened until the 1960s.

After service
CITY OF GLASGOW was sold to the Admiralty in December 1917 after the Navy contacted the RNLI requesting her use. Although the Admiralty was not seeking to take over lifeboats, the steam lifeboat was an exception. As a result, the RNLI suggested they purchase the lifeboat with the option for the Institution to reacquire her at the end of the war. In December 1917 the Admiralty decided to go ahead with the acquisition and bought the boat and her equipment for £4,290, subsequently converting her into a patrol boat named Patrick. A service carried out by Patrick when she was under Admiralty control involved the German submarine UC-1, a minelayer, which hit one of its own mines off the Sunk Lightship on 26 June 1918. A huge explosion was seen by Patrick's crew, and they reached the scene to find a lone survivor, the commander of UC-11, Oblt. Kurt Utke, floating nearby. The vessel had hit its own mine while at periscope depth. It is believed that Patrick was later sent to the Nile, but further details of her whereabouts and career are not known.

ON.446

▲ City of Glasgow at moorings in the Pound at Harwich. She carried auxiliary sails in case of engine failure, comprising a forelug and jib. (By courtesy of Harwich RNLI)

▲ A fine profile view of the second City of Glasgow at her usual moorings in the Pound at Harwich. She was the last steam lifeboat to be built by the RNLI. (By courtesy of the RNLI)

STEAM LIFEBOATS

Helen Peele

Key data
BUILDER 1901, Ramage & Ferguson Ltd, Leith
DONOR Legacy of Mr C. J. Peele, Chertsey
NAMED 28 Jun 1901 at Leith
COST £9,784 10s 0d
DIMENSIONS 95ft 6in x 19ft 6in x 10ft 7in; draught 7ft 9in forward, 9ft 3in aft
TONNAGE 133 gross tons, 16 net
STATIONS Padstow Sep 1901 – Aug 1917 and May 1919 – May 1929
RECORD 24 launches, 20 lives saved (totals for services performed on her own)
DISPOSAL Sold 1 May 1929 for £950 to Captain John Turner

Service at Padstow
HELEN PEELE was a unique steam tug designed by G.L. Watson and purpose-built by the RNLI for service off the coast of North Conwall. Her two compound steam engines, each with two cylinders, produced 331ihp, running at 191rpm, with a steam pressure of 104lbs, giving a speed of 10.4 knots. She was sent to Padstow on 11 September 1901, and worked in conjunction with the lifeboats in the area and, apart from a spell with the Admiralty during the First World War, she spent her entire career in Cornwall. Although mainly used for towing the 42ft twelve-oared self-righting lifeboat Edmund Harvey to wrecks, on at least one occasion she towed the station's smaller lifeboat Arab, and on another worked with the St Ives lifeboat. The Admiralty took over the tug, along with her officers and crew, in August 1917 and she was initially sent to Portland. She was later used for towing duties in Shetland, as there was no lifeboat at Lerwick, and assisting small ships in distress. She continued in service at Lerwick until May 1919, and was then returned to Padstow.

After service
HELEN PEELE was sold out of service in May 1929 to Captain John Turner. She was converted into a private yacht in August 1929, being fitted with accommodation including a saloon, two double staterooms, two single staterooms, two bathrooms, and accommodation forward for a crew of six, and her original wheelhouse was converted into a deck lounge. She was later owned by Lord Marchamley, London and in 1932 by Captain Thomas W.F. Winter, of Sevenoaks, Kent. Little is known of her subsequent career and she was last reported as being employed as a tender on the Clyde in March 1964.

ON.478

▲ The steam tug Helen Peele in Padstow harbour. The impressive 95ft 6in twin-screw vessel was designed by George Lennox Watson and represented a unique departure for the RNLI. (By courtesy of Padstow RNLI)

▲ Helen Peele taking local lifeboat supporters for a trip in the Camel Estuary. (By courtesy of Padstow RNLI)

STEAM LIFEBOATS

Steam lifeboats around the world

AS WELL AS THE RNLI'S SIX STEAM lifeboats, a number of other such craft were built for service elsewhere. Generally speaking, those built for Commonwealth countries were expected to perform other duties as well as lifesaving, such as the boarding and disembarking of pilots. Perhaps surprisingly, the two built for Australia remain in existence.

CITY OF ADELAIDE As already discussed, the Koninklijke Zuid-Hollandsche Maatschappij tot Redding van Schipbreukelingen (KZHMRS, South Holland Lifeboat Society) ordered a steam lifeboat from R. & H. Green at Blackwall, named President van Heel, but construction delays resulted in the Dutch refusing to take delivery of the boat. As a result, the South Australian Government purchased the craft, funded by a gift of £3,500 from Robert Barr Smith, of Torrens Park, co-founder of Elder, Smith & Co.

She was an almost exact replica of the first City of Glasgow, measuring 52ft in length by 15ft 2in in beam, with a displacement of thirty tons, and maybe the reliability issues which plagued City of Glasgow influenced the Dutch Society's decision to refuse the boat. She had a single 200ihp compound steam engine driving two centrifugal pumps on a horizontal shaft. The pumps supplied water to nozzles that gave her a maximum speed of eight knots.

The boat was renamed City of Adelaide and arrived in South Australia in 1896, when she was formally commissioned by Smith and presented to the South Australian government for service at Beachport, where she stayed until 1930, but amazingly was never called upon for her primary use. She could accommodate up to eight crew and forty survivors, had a maximum speed of eight knots and a range of 200 nautical miles. The steam-powered water jets were not a success, however, and after City of

▲ City of Adelaide on display at the Axel Stenross Maritime Museum at Port Lincoln. Although badly deteriorated, the craft is an unusual survivor from the days of steam power. (Heritage South Australia)

▶ Lady Forrest was built in 1903 as a steam-powered pilot boat-cum-lifeboat and served Fremantle Port Authority for more than sixty years.

Adelaide was commissioned at Glenelg, she had to be towed on her delivery trip to Beachport because the water inlets became blocked by weed.

After just over a decade, the steam plant was completely replaced in March 1909 with a 30kw (40hp) petrol engine at a cost of £510 by Mort's Dock and Engine Company of Sydney. The boiler had already been replaced at considerable expense, but despite this the power developed had never been sufficient to drive the boat, and she could rarely manage more than four knots. In the space where the boiler had been, four bunks were the fitted, and there was then enough room for about forty survivors. In 1911 the water jets and pumps were replaced by a propeller.

When the lifeboat service was disbanded, City of Adelaide was sold and she was used by the Salvation Army as a fishing launch. The next owner, Benno Hage, used it for fishing, picnics and carrying sheep, wool and firewood. She was operated by his son Dudley and a four-cylinder petrol tractor engine was fitted around this period. After about three years Hage sold City of Adelaide to Roly Puckridge who fitted a petrol and kerosene Fordson engine alongside the existing one, with a chain drive linking the two. He maintained the craft for the same mixed use as the previous owners.

In 1952 the boat ended her seagoing days and was sold to boat builder Axel Stenross, who removed all the fittings. The hulk then lay on the beach at Porter Bay, Port Lincoln for many years, until in the late 1970s she was declared an historic shipwreck. In 1985 Port Lincoln Council decided to remove the hulk to the Axel Stenross Maritime Museum, where it now remains on display to the public.

LADY FORREST Another steam lifeboat was built for service in Australia at the

STEAM LIFEBOATS

Steam lifeboats around the world (continued)

◀ Fremantle Harbour Trust pilot boat Lady Forrest in 1964. She was converted to diesel power in the 1940s and a superstructure fabricated from the conning tower of the Netherlands submarine was added.

start of the twentieth century. Named Lady Forrest, she was designed by H. Douglas and built in 1903 by J.S. White at Cowes. She was a screw vessel, a sister to the second City of Glasgow, and measured 56ft overall. She was built to the order of the Fremantle Harbour Trust (later the Fremantle Port Authority) after several ships were wrecked off the port, including the barques Carlisle Castle, with the loss of all her crew, and City of York, both in July 1899. Both losses were attributed to an inadequate service provided by the Fremantle pilot boat, and Lady Forrest was sent to improve the situation.

The new steam boat left London on 26 May 1903 as deck cargo on board Scottish Shire Line's cargo steamer Fifeshire and, on 11 July 1903, arrived at Fremantle where she was put in the water on 14 July. After further fitting out, which took until 25 July, she was ready for service and her first task was to take three pilots to the pilot station at Rottnest Island. She took just over an hour to accomplish the twelve-mile crossing, and returned in an hour with the benefit of a breeze. Her final acceptance trial was on 27 July and she then settled down to her pilotage duties, also going on various occasions in search of vessels in difficulties, making a number of rescues.

Named after the wife of the first Premier of Western Australia, Lady Forrest gained a fine record over more than sixty years helping vessels in all weathers. In about 1946 she was converted to diesel power and rebuilt with a superstructure fabricated from the conning tower of the Netherlands submarine K.15, which had been scrapped locally after wartime damage.

In 1959 she was replaced by Lady Gairdner, and she became a stand-by pilot vessel. She was officially retired and lifted from the water on 23 June 1967 to be rebuilt to her original form, the work being completed on 6 April 1968. On 25 November 1970 she was incorporated into the WA Maritime Museum, Fremantle.

Princess of Wales at Chiswick, on the Thames, shortly after she had been completed by Thornycroft.

PRINCESS OF WALES A steam lifeboat-cum-pilot vessel, Princess of Wales was built by Thornycroft and Co at Chiswick in 1903, for the Crown Agents for the Colonies, a quasi-governmental organisation founded in 1833 to administer the British Empire. The boat was intended for service in Mauritius. She was screw-propelled and was the largest of any steam lifeboat, measuring 59ft 10½in in length. Her engine was of 142hp and she was capable of speeds up to 9.65 knots. In appearance she was remarkably similar to her RNLI sisters with twin funnels, a mast abaft the funnels and an open cockpit aft.

MOLESEY The second steam lifeboat-cum-pilot vessel built for the Colonies was Molesey, of Lagos, which was outwardly similar to the RNLI screw lifeboats but had the benefit of twin screws. She was completed in 1905 by Forrestt and Co, at their Wivenhoe yard. Her waterline length was 57ft, breadth 12ft 6in (extreme 15ft) and she had a laden draught of 3ft 6in. Her twin engines, supplied by a Yarrow watertube type of boiler, gave her just over ten knots on trial, rather more than the contracted speed of nine and a half knots.

The 1905-built steam lifeboat Molesey.

STEAM LIFEBOATS

Steam lifeboats in the Netherlands

The ZHMRS steam lifeboat President van Heel served at Hoek van Holland from 1895 to 1930. Her first rescue was completed on 18 June 1896, when she saved five from a stranded English schooner. In 1901 she brought thirty-seven people ashore from the stranded Italian steamship Bakio.

PRESIDENT VAN HEEL The Zuid-Hollandsche Maatschappij tot Redding van Schipbreukelingen (ZHMRS, South Holland Lifeboat Society) built and operated two steam lifeboats, both of which were stationed at the important Hook of Holland station, which was situated at the entrance to the River Maas. As Rotterdam was becoming the busiest port in the world, such a lifeboat was deemed essential to cover the port's entrance.

The ZHMRS decided in 1893 to obtain a steam lifeboat and ordered a boat from R. & H. Green in Blackwall, with construction to start in June 1893. Penn and Sons were to supply the steam engine and waterjet propulsion, including a water pump capable of pumping one ton of water per second, providing propellerless propulsion in any desired direction. The total cost was £3,500. Due to various setbacks, however, the completion date was repeatedly delayed and the ZHMRS ran out of patience and cancelled the contract. As described above, this boat was completed for the South Australian Government.

After a delay of more than a year, a new contract was signed, this time with John I. Thornycroft & Co at Chiswick. Construction went well and, after successful sea trials on the Thames, the new President van Heel was commissioned on 25 September 1895, entering service soon afterwards. The boat, which cost £4,750, was named after ZHMRS chairman J.J. Marie van Heel, who had been a committee member since 1857 and had already paid for two lifeboats.

Designed by G.L. Watson, the new boat

◀ President van Heel shows her speed at sea off the Hook; ▲ and at the boatyard following her capsize in October 1921. (By courtesy of Hoek v Holland KNRM)

was similar in external appearance to Duke of Northumberland, with twin funnels, a mast forward and an open cockpit aft, a length of 55ft (16.5m) overall, and 15ft 9in (4.8m) in beam. She was fitted with the builder's own patent watertube boiler and a compound engine developing 125hp, driving a single propeller of 30in in diameter. On trials, President van Heel reached a speed of 9.29 knots. She had a displacement of 21 tons.

President van Heel served at Hook of Holland for more than thirty years, undertaking many rescues. However, on 24 October 1921 she capsized while on service to the French barge Falaise. Six of her crew of seven were drowned and she was washed ashore on the sands, upside down, near the South Pier.

President van Heel at her berth after the tragic Berlin shipwreck in 1907, when her crew took her out to help. (By courtesy of Iain Booth)

STEAM LIFEBOATS

Steam lifeboats in the Netherlands (continued)

▶ President van Heel pictured in 1922 after she had returned to service following her capsize the previous year. The six crew who died in the capsize were J.P. de Geus (skipper), C.P. Weenig, D. Boel, B.W. Boxman, W. Van der Klooster and A. Visser.

She was repaired and returned to service in 1922, serving a further eight years before being sold. She was later converted into a fishing vessel and then a pleasure boat, being kept usually in Rotterdam.

PRINS DER NEDERLANDEN The construction of a second steam lifeboat came about following the wreck of the Great Eastern Railway Company's steamer Berlin (built 1894) at the Hook on 21 February 1907. President van Heel attempted to help, but rough seas prevented her from approaching the stricken vessel. The pulling lifeboat also performed valuable services in appalling conditions, but was unable to help and Berlin broke in two amidships. Desperate efforts were made to save those on board, but 128 of 148 persons were drowned.

Following this, the ZHMRS resolved to obtain a second steam lifeboat. Although the RNLI had by then turned to screw propulsion, and the North Holland Lifeboat Society had built a motor lifeboat in 1907, the ZHMRS were so satisfied with the waterjet propulsion of President van Heel that they ordered a replica from the Fijenoord Yard at Schiedam, Rotterdam.

Named Prins der Nederlanden after Prince Hendrik, patron of the ZHMRS, she was sent to the Hook in 1909 and served the station for twenty years. In 1924 she visited London, as one of several overseas lifeboats attending the celebration to mark the centenary of the foundation of the RNLI and was demonstrated to delegates at the first International Lifeboat Conference.

Her career ended on 16 January 1929, when, on a cold winter's day with a biting

▲ Prins der Nederlanden at the berth in Hook of Holland from where the lifeboats were operated, with crew and officials on board. (By courtesy of KNRM)

north wind and snow showers, she went to the aid of the Latvian freighter Valka, which had gone aground on the Maasvlakte sandbank. Arriving on scene, the lifeboat crew attempted to get a line aboard the casualty, but failed and the lifeboat was repeatedly slammed against the side of the ship. As a third attempt was made, the lifeboat was caught in the surf and capsized, being washed ashore to the south of the Maas. Those on board the stranded Valka, twenty-five in all, were rescued the next day by another lifeboat, but Prins der Nederlanden was scrapped following the tragedy.

▲ Prins der Nederlanden upturned on the Goereese coast in January 1929. The entire crew were lost in the disaster: A. v.d. Klooster (Coxswain), H.P. Meyboom, J.M. Timmers, P. van Asperen, R. de Groot, A. Meuldijk, C. Sterrenburg and P.A. Verwey. (By courtesy of KNRM)

STEAM LIFEBOATS

Stations served by RNLI steam lifeboats

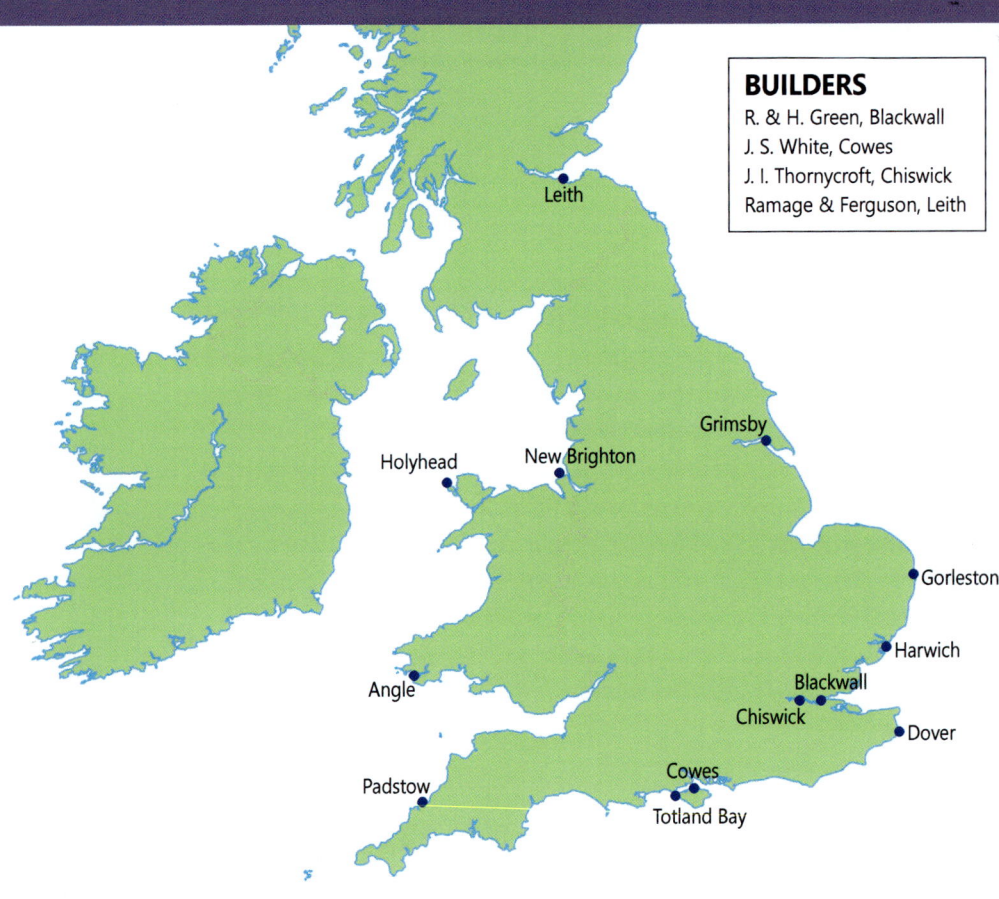

BUILDERS
R. & H. Green, Blackwall
J. S. White, Cowes
J. I. Thornycroft, Chiswick
Ramage & Ferguson, Leith

THE STEAM LIFEBOATS AND THEIR YEARS OF SERVICE

1 • Duke of Northumberland 1890-1922
2 • City of Glasgow 1894-1901
3 • Queen 1897-1923
4 • James Stevens No.3 1898-1928
5 • James Stevens No.4 1899-1900
6 • City of Glasgow 1901-1917
7 • Helen Peele (tug) 1901-1929

STATIONS SERVED BY STEAM LIFEBOATS

Angle
(4) 1908-1915
Dover
(4) 1919-1922
Gorleston
(2) 1897-1898
(4) 1903-1908
Grimsby
(4) 1898-1903

Harwich
(1) 1890-1892
(2) 1894-1897 and 1898-1901
(6) 1901-1917
Holyhead
(1) 1892-1893 and 1897-1922
(4) 1922-1928

New Brighton
(1) 1893-1897
(3) 1897-1923
Padstow
(5) 1899-1900
Totland Bay
(4) 1915-1919